STUDY GUIDE

Fourth Edition

Michelle Calvarese
California State University, Fresno

THE POWER OF PLACE

Geography for the 21st Century

The Power of Place: Geography for the 21st Century is produced by Cambridge Studios in collaboration with Australian Broadcasting Corporation, Open Learning Australia; Centre National de Documentation Pédagogique, France; NHK School Broadcasts Division, Japan; the Teleac Foundation, the Netherlands; and Swedish Educational Broadcasting Company. Funding is provided by Annenberg/CPB.

McGraw Hill Higher Education

Boston Burr Ridge, IL Dubuque, IA Madison, WI New York San Francisco St. Louis
Bangkok Bogotá Caracas Kuala Lumpur Lisbon London Madrid Mexico City
Milan Montreal New Delhi Santiago Seoul Singapore Sydney Taipei Toronto

The McGraw·Hill Companies

Front cover image: Han Chinese and Chinese Muslim school children perform a daily exercise routine.

Study Guide, Fourth Edition by Michelle Calvarese
THE POWER OF PLACE: GEOGRAPHY FOR THE 21ST CENTURY

Published by McGraw-Hill Higher Education, an imprint of The McGraw-Hill Companies, Inc., 1221 Avenue of the Americas, New York, NY 10020.

This book was developed for use with **The Power of Place: Geography for the 21st Century**, an Annenberg/CPB project video series and distance learning course. The course consists of 26 half-hour video programs; the McGraw-Hill textbook *Contemporary World Regional Geography: Global Connections, Local Voices*, by Bradshaw/White/Dymond; a Faculty Guide; and this Study Guide.

The Power of Place: Geography for the 21st Century is produced by Cambridge Studios in collaboration with Australian Broadcasting Corporation, Open Learning Australia; Centre National de Documentation Pédagogique, France; NHK School Broadcasts Division, Japan; The Teleac Foundation, the Netherlands; and Swedish Educational Broadcasting Company.

Arrangements for Use
Colleges, universities, and other educational institutions may:

- purchase the programs on videocassette for use in the classroom and media center (1/2" VHS cassettes); $499 for the series. For more information call 1-800-LEARNER or order online at www.learner.org.
- duplicate the programs for use in the classroom or media center. For information call 1-800-LEARNER.
- license the use of **The Power of Place** as a course for distant learners or acquire an off-air or off-satellite taping license through the PBS Adult Learning Service. For information call 1-800-257-2578 or e-mail als.pbs.org.

The Annenberg/CPB Project
Annenberg/CPB, a partnership between the Annenberg Foundation and the Corporation for Public Broadcasting, uses media and telecommunications to advance excellent teaching in American Schools. Annenberg/CPB funds educational series and teacher professional development workshops for the Annenberg/CPB Channel. The Channel is distributed free by satellite to schools and other educational and community organizations nationwide and streamed via broadband on the Web. Funding for **The Power of Place: Geography for the 21st Century** is provided by Annenberg/CPB.

This book is printed on acid-free paper.

2 3 4 5 6 7 8 9 0 BKM BKM 0 9 8 7 6 5

ISBN 0-07-281855-7

www.mhhe.com

Contents

Course Preface

LESSONS IN GEOGRAPHY

At the end of **The Power of Place: Geography for the 21st Century** you should be able to engage in creative and insightful discussion about these questions:

1. How and why are phenomena arranged in particular ways on the earth's surface?
2. What are the factors that create or change regions, places, environments, and landscapes?
3. How are certain events and processes shaped by human/environment interactions?
4. Describe the physical and human patterns found in the Case Study sites.
5. How is each Case Study site distinguished by a link between the human and physical world?

COURSE OVERVIEW

The course you are about to explore, **The Power of Place: Geography for the 21st Century**, is the culmination of an unusual collaboration by an international team of educational broadcasters and geographers from the United States, Australia, France, Japan, the Netherlands, and Sweden. Evolving from a concern in each country that its educational system promote greater understanding about the Earth's environments and its peoples, this course provides a global perspective on the subject of World Regional Geography.

The discipline of Geography is uniquely suited to address the world's human and environmental challenges because it is a field of study distinguished by a focus on the location and distribution of human and physical activity across the face of the Earth, the interaction of people and environment in particular locations, and the distinctive places and regions that result. One of the striking features of Geography is that little distinction is made between domestic and international matters. Such categorization of the field, in fact, has an odd ring to the professional geographer. By embracing a field of study whose goal since the time of the Greeks has been to "write about the Earth," the geographer accepts the challenge of understanding and explaining the myriad forces that shape the world's diverse physical and social environments.

The Power of Place: Geography for the 21st Century embraces this challenge. The course provides an examination of many geographic forces currently shaping the complex features of the world's civilizations and environments. The course includes 26 half-hour Video Programs—organized into 10 Units—providing insights into a range of geographic issues around the globe. All told, the course visits 34 countries and over 50 sites around the world. The series commences with two half-hour Introductory Programs, presenting an overview of the course and highlighting the geographic concepts around which the project is organized. The remaining 24 programs that comprise **The Power of Place** each contain two 12-minute documentary style case studies; the course is linked together with the companion textbook *Contemporary World Regional Geography: Global Connections, Local Voices* by Bradshaw/White/Dymond, © 2004 by The McGraw-Hill Publishing Companies.

COURSE OBJECTIVES

To encourage the study of geography, we have used video technology to provide the feel of a place, something difficult to reproduce in other media. Each Case Study focuses on people whose lives are shaped by particular geographical forces. In addition, we explore a vital regional and conceptual issue that can be further illuminated through geographic analysis. Each program, then, is designed to fulfill two major goals: (1) to characterize selected examples of regions within the 9 geographic regions explored, and (2) to communicate one or more important concepts from a variety of topical approaches—for example, the fields of physical, political, historical, economic, and cultural geography.

Course Study Guides are designed and published for the purpose of linking the video series and its companion text. This 4th edition of **The Power of Place** *Study Guide* has been rewritten to accompany the new companion text for this course: *Contemporary World Regional Geography: Global Connections, Local Voices,* by Bradshaw/White/Dymond, © 2004 by The McGraw-Hill Publishing Companies.

In order to facilitate student learning during the course, students should familiarize themselves with the following study resources.

COURSE STUDY RESOURCES

1. The 26 half-hour Video Programs.
2. **The Power of Place: Geography for the 21st Century** *Study Guide* by Michelle Calvarese, © 2004 by The McGraw-Hill Publishing Companies.
3. *Contemporary World Regional Geography: Global Connections, Local Voices*, by Bradshaw/White/Dymond, © 2004 by The McGraw-Hill Publishing Companies.
4. **The Power of Place: Geography for the 21st Century** Coordinated Website: www.learner.org/powerofplace, copyright Annenberg/CPB.
5. *Teaching Geography* Coordinated Website: www.learner.org/channel/workshops/geography, copyright Annenberg/CPB.

THE COURSE UNITS

UNIT 1: GLOBALIZATION AND WORLD REGIONS
Program 1: One Earth, Many Scales

Lost in Space? Geography Training for Astronauts

A NASA geographer trains Space Shuttle astronauts to recognize and photograph places on Earth. Introduces key issues in physical geography and human-environmental interaction.

Globalization and Revolt

Why do the forces of globalization seem to draw some places closer together and cause others to pull farther apart?

Program 2: Boundaries and Borderlands

Twin Cities, Divided Lives

To support her family, a single Mexican mother illegally smuggles cigarettes—into Texas. Her daily struggle for survival introduces us to concepts of relative location and geographic regions.

Operation Hold the Line

The U.S.-Mexico borderlands form a unified cultural and economic region with qualities of both states. But efforts to reduce illegal border crossings have unintended consequences at a global scale.

UNIT 2: EUROPE
Program 3: Supranationalism and Devolution

Strasbourg: Symbol of a United Europe

As the seat of the European Union's Parliament and the Council of Europe, the city of Strasbourg is one locus of power in an increasingly supranationalist Europe.

Slovakia: New Sovereignty

In the 1990s Czechoslovakia separated into the Czech Republic and the Slovak Republic. How has sovereignty affected the people of Slovakia?

Program 4: East Looks West

Berlin: United We Stand

Berlin in now capital of a reunified Germany and a symbol of a more unified Europe. But the integration of East Berlin requires urban reorganization and economic expansion.

Poland: Diffusion of Democracy

Strategies for spreading democracy through Poland required a decidedly spatial approach, focusing on local actions in rural areas.

Program 5: The Transforming Industrial Heartland

Liverpool: A Tale of Two Cities

Arts and service industries follow EU investment into the old port city. Will it turn the tide of economic decline in Merseyside?

Randstad: Preserving the Green Heart

Small, crowded Netherlands operates the largest port in the world and the fourth largest airport in Europe. Can they maintain their transportation advantage and also preserve their dwindling green space?

Program 6: Challenges in the Hinterlands

Andalucia: Developments in the Hinterlands

At the southern end of Southern European Spain, Andalucia has struggled to integrate with the heartland. Can they move beyond tourism and agriculture?

Iceland: Edge of the Habitable World

At the borders of the habitable world, Iceland learns the lessons of sustainable fish harvests. But can it manage the social costs?

UNIT 3: RUSSIA AND NEIGHBORING COUNTRIES

Program 7: Northwest Contrast

St. Petersburg: Russia's Window on the West

Czar Peter the Great built St. Petersburg to strengthen connections to Europe. What challenges continue to face this Russian port in this post-Soviet society?

Vologda: Russian Farming in Flux

Under the Soviets, Russian farms were state-owned collectives. But with the move to privatization, has anything really changed?

Program 8: Holding the Hinterlands

Dagestan: Caucuses Disconnect?

In the Russian republics of the North Caucuses, Chechen rebels fight for independence. The question of national stability may rest in the neighboring Islamic republic of Dagestan.

Bratsk: The Legacy of Central Planning

Communist ambitions create the world's largest hydroelectric project followed by a Russian city in the middle of Siberia.

UNIT 4: EAST ASIA

Program 9: Changes on the Chang Jiang

Shanghai: Head of the Dragon

After years of stagnation, Shanghai and its suburbs enter the 21st century on a wave of development, ready to reclaim the legacy as China's commercial center for decades to come.

Sijia: Small Town, Big Change

On the outskirts of urban Nanjing, the steady growth of a township enterprise illustrates three great contrasts in modern China: rural vs. urban; agricultural vs. industrial; coastal vs. interior.

Program 10: The Booming Maritime Edge

Guangdong: Globalization in the Pearl River Delta

This program explores globalization in East Asia. In Guangdong, we see the effects of modernization on Chinese society.

Taiwan: High-Tech Tiger

What factors contributed to Taiwan's emergence as a high-tech powerhouse? What role can Taiwan play in China's developing high-tech industry?

Program 11: A Challenge for Two Old Cities

Lanzhou: Confluence of Cultures

To understand the balance of power between two major cultures, we travel to the frontiers of Han and Muslim China in the city of Lanzhou.

Shenyang: Hope for China's Rust Belt?

In northern China, this previously dynamic industrial city continues to struggle with modernizing its manufacturing infrastructure.

Program 12: Small Farms, Big Cities

Northern Japan: Protecting the Harvest

Rice farmers battle destructive weather in order to save their crops. This is a fascinating look at human-environmental interaction in East Asia.

Tokyo: Anatomy of a Mega-City

The continuing expansion of the Tokyo megalopolis leads to ever longer commutes and demand for suburban housing.

UNIT 5: SOUTHEAST ASIA AND SOUTH PACIFIC

Program 13: The Mainland

Laos: Isolated Heart

Laos is emerging from isolation to join the global economy as an exporter of hydroelectric power. Today it tries to balance economic development with environmental protection.

Vietnam: Fertile Dreams

Environmental factors and improved agricultural technique have made Vietnam the world's second largest rice exporter. Its booming economy is also evident in the explosive growth of Ho Chi Minh City.

Program 14: The Maritime Connection

Indonesia: Tourist Invasion

Their culture once imperiled by hordes of tourists, Balinese residents have developed strategies to profit from the tourist industry while maintaining cultural Integrity.

Multi-Cultural Malaysia

Amidst growing pressures from Islamic militants, social and economic programs seek to build tolerance among Malaysia's diverse cultures.

Program 15: Global Interaction

Singapore: Gateway to Southeast Asia

High-tech infrastructure, a well-educated workforce and strict government repression have all helped Singapore achieve its status as one of the pre-eminent ports and wealthiest cities in the world.

Australia: New Links to Asia

Abundant in the natural resources and wide-open spaces, Australia has long been an exporter of raw commodities to Europe. Recently, links have shifted to Asian economic tigers in their own region.

UNIT 6: SOUTH ASIA

Program 16: Urban and Rural Contrasts

Delhi: Bursting at the Seams

Delhi, the ever-expanding capital of India continues to act as a magnet, pulling millions of Indians away from the hardships of the rural countryside.

Dikhatpura: Help through Irrigation

Although government programs encouraging irrigation and modern agricultural techniques have improved life for many people, creating sustainable development proves a challenging proposition.

UNIT 7: NORTHERN AFRICA AND SOUTHWESTERN ASIA

Program 17: Sacred Space, Secular States?

Jerusalem: Capital of Two States?

The old city in Jerusalem contains sacred spaces of three religions. It is also the symbolic focus of a nationalist conflict between Palestinians and Jews. Can the historical and political geography of this holy city provide clues to a peaceful resolution?

Turkey: Fundamental Change

At the edge of Europe, Turkey hopes to take economic advantage of its proximity to the western world. Yet as the power of Islam grows, this country may face potentially divisive fundamentalist forces.

Program 18: Oil and Water

Egypt: Gift of the Nile

This program investigates Egypt's limited natural resources, focusing on that nation's dependence on the Nile River.

Oman: Looking Beyond Oil

Having benefited greatly from its relatively modest oil reserves, Oman looks to diversify its economy for future growth. One of its steps is unusual for an Islamic country—promoting the education and employment of women.

UNIT 8: AFRICA SOUTH OF SAHARA

Program 19: Strength to Overcome

South Africa: This Land Is My Land

South Africa continues to face many challenges in redressing the land inequities under apartheid. Join a geographer researching one type of land reform and its results.

Kenya: Medical Geography

AIDS has become one of the biggest killers in Kenya. Geography can play an important role in understanding how this and many tropical diseases spread. What are the developing strategies to reduce loss of life?

Program 20: Developing Countries

Cote d'Ivoire: Cocoa and Change

Cote d'Ivoire gained its independence from France in 1960. Since then it has been the world's largest producer of cocoa, but has also experienced economic downturns and loss of its historically stable government.

Gabon: Sustainable Resources?

In one of Africa's wealthiest countries, oil revenues have declined, putting new pressure on the country's timber resources. Working with U.S. conservation organizations, the president puts aside 10% of Gabon's rainforest in preserves.

UNIT 9: LATIN AMERICA

Program 21: Population Geography

Mexico: Motive to Migrate

A geographer's research reveals a major source of Mexican migration: the North-Central 'Hollow Core.' Will southward diffusion of maquiladoras or a new U.S. border policy have a greater impact keeping people home?

Guatemala: Population and Conquest

Every year a greater number of Maya Indian victims of 'continuing conquest' must share inadequate agricultural resources as their growing population outstrips the carrying capacity of their meager land.

Program 22: The Dynamic Pacific Rim

Ecuador: Orange Alert

Scientists monitoring the Tungurahua Volcano suddenly see dangerous signs and have to advise the government: do they recommend evacuation and ruin the local economy or wait and risk many lives?

Chile: Pacific Rim Player

Bordered to the east by the towering Andes Mountains, and on the west by the Pacific Ocean, Chile enjoys continued economic growth, solidifying its position as a Pacific Rim player.

Program 23: Brazil: The Sleeping Giant

Sao Paulo: The Outer Ring

The sprawling mega-city of Sao Paulo is evidence that Latin America is among the most rapidly urbanizing regions. There, a constant stream of new arrivals live in rings of squatter settlements along the edges of this ever-expanding metropolis.

A Second Chance for Amazonia?

An American scientist discovers new possibilities for sustainable development as the forces of civilization push deforestation deeper into the Amazon basin.

UNIT 10: NORTH AMERICA

Program 24: Cityscapes, Suburban Sprawl

Boston: Ethnic Mosaic

After proposing a federal empowerment zone in Boston's diverse but poor neighborhoods, the city initially wins 'second place.' What do they accomplish with those funds before they finally win the grand prize?

Chicago: Farming on the Edge

As in many areas of the U.S., suburban Chicago just keeps expanding into the surrounding countryside. The big difference here: they are losing perhaps the world's richest soils.

Program 25: Ethnic Fragmentation in Canada

Vancouver: Hong Kong East

Prior to the Chinese takeover of Hong Kong, thousands of wealthy businessmen move their families to Vancouver causing a collision of cultures. What has happened since 1997?

Montreal: An Island of French

Trying to preserve their culture, Quebec welcomes immigrants and pays to teach them French. A referendum on Quebec sovereignty fails by a narrow margin.

Program 26: Regions and Economies

Oregon: A Fight for Water

Native Americans and farmers compete for a scarce resource: water. Can the two sides reconcile their conflict and restore Pacific salmon to one tributary of the Columbia River?

U.S. Midwest: Spatial Innovations

The U.S. Midwest has seen a tremendous change in the past 30 years. Here, an influx of Japanese automakers has brought more than just new factories to this once declining manufacturing region.

THE POWER OF PLACE STUDY GUIDE

The purpose of this *Study Guide* is to aid students in their study of the material presented in **The Power of Place.** Each Unit of this course explores a specific region of the world and analyzes it through one or more 30-minute Programs consisting of two Case Studies. The *Study Guide* should be consulted before viewing each Unit and Program, to help students draw out these important aspects of the course:

- **Unit Objectives** target the learning goals for the programs within each Unit.
- **Key Themes** are listed, so students may watch for and identify these as they view each Program.
- **Program overviews** appear after the key themes as brief narratives, to help direct student focus toward the main points of each Program.

After viewing the Programs within a unit, students may again consult the *Study Guide* to help integrate the themes and test their understanding of the materials:

- A Discussion of Unit Themes explains in more detail the unit objectives the students read before viewing the Programs, and makes clear why these are important themes. This section of the *Study Guide* also provides specific references to places in the companion textbook, *Contemporary World Regional Geography* by Bradshaw/White/Dymond, where students can read additional information on the specific region or concept being covered in the course Unit.
- The **Test Your Knowledge** sections provide questions to test the most important concepts and facts from the Programs within the Unit through a variety of multiple choice, true/false, short answer, and essay questions.
- **Map exercises** help students to identify places, borders, physical features, and regions on the outline maps.

WHAT IS A TELEVISION COURSE?

The primary difference between television courses and traditionally taught courses is the manner in which instruction is delivered to students. Traditional students may come to campus for direct instruction by a faculty member several times per week; television course students work more independently, watching the television programs and reading the print materials at home or work with guidance from the course faculty through a variety of communication and instructional techniques. The number of required class meetings, however, is generally much fewer than in traditionally taught classes because the basic course content is delivered to students through television, print, and other types of faculty/student interactions.

Television courses, in most other respects, e.g., academic rigor, student requirements, and need for qualified faculty, are equivalent to traditionally taught college courses. Television course students enroll in the college or university that has adopted the course, pay tuition to that institution, usually have access to student services offered at that institution, receive academic credit from that institution, and are supervised by that institution's faculty. These faculty members make similar kinds of academic and instructional decisions as they would in traditionally taught classes, and interact with students through a combination of classes, written assignments, phone contacts, and mailings.

TAKING THE COURSE

Find Out the Following Information When You Register:

- What books are required for the course?
- Are orientation sessions scheduled?
- When will **The Power of Place** be broadcast in your area?
- How will your college arrange to get you copies of the video programs?
- When are course examinations scheduled?
- Are any on-campus meetings scheduled?

Be an Active Learner

1. Before viewing each Video Program, read the *Study Guide*'s **Unit Objectives** for the Unit in which your Program is located.
2. Read the textbook assignment listed in the **Discussion of Unit Themes** section and any supplementary material assigned by your instructor. Note any **overviews**, lists of **Key Themes**, and **summaries** that will help you identify important information.
3. View the Program, keeping the **Unit Objectives** and **Key Themes** in mind. Be an active viewer. You will want to record the program on a VCR if you own one. Some public television stations also repeat programs during the week, or your college may arrange a local cable broadcast. Some students find it helpful to take notes while viewing the program.

4. After viewing the Program, read the **Discussion of Unit Themes** in the *Study Guide*.
5. Use the **Test Your Knowledge** sections within the *Study Guide* to be sure you understand the concepts, themes, and place locations discussed.
6. Keep up with the course on a weekly basis. Each Unit of the course builds on knowledge gained in previous Units. Stay current with the programs and readings. Make a daily checklist and keep weekly and term calendars, noting scheduled activities such as meetings or examinations, and set aside blocks of time for viewing programs, reading, and completing assignments.
7. Keep in touch with your instructor. If possible, get to know him or her. You should have your instructor's mailing address, phone number, and call-in or office hours. Your instructor would like to hear from you and to know how you are doing, and he or she will be eager to answer any questions you have about the course.
8. Utilize these additional online geography study tools for further course enrichment:

   Teaching Geography Coordinated Website: www.learner.org/workshops/geography
   Complementing the *Teaching Geography* workshops, this rich resource provides further connection to the national geography standards, workshop activities, lesson plans, interactive maps and activities, and additional links to content resources.

   The Power of Place Coordinated Website: www.learner.org/powerofplace
   This companion site connects case study content to the national geography standards, highlights keywords and concepts, provides links to web- and text-based resources, and includes further exploration of issues raised by each case study.

UNIT ONE: GLOBALIZATION AND WORLD REGIONS

Unit Objectives

At the completion of the unit, you should be able to:

- understand the importance of globalization.
- understand the spatial perspective of geography.
- understand the functions of regions.

PROGRAM 1

ONE EARTH, MANY SCALES

Key Themes

- **Globalization**
- **Scale**
- **Spatial perspective**
- **Human-environmental interaction**

Lost in Space? Geography Training for Astronauts

Astronauts rely on a spatial perspective of the earth each time they lift off into space. Images from space can give us valuable information about the physical properties of the earth's surface. Satellites circle the earth taking snapshots, which can later be fitted together to provide a comprehensive image of the earth's topography, physical systems, as well as dangers and limitations. Night views from space can tell us about population patterns and economic development. Environmental degradation in the form of desertification and air pollution is also visible from space.

Globalization and Revolt

In light of recent events, the notion of globalization has become increasingly important. The world is becoming more and more interconnected and understanding these connections is imperative to understanding how the world works. This interconnectivity expresses itself in many diverse forms such as fashion trends, technology, the cultural landscape, and terrorism. Geography can analyze these connections on both a global and local scale.

PROGRAM 2

BOUNDARIES AND BORDERLANDS

Key Themes

- **Location**
- **Region**
- **Scale**

Twin Cities, Divided Lines

Once a barren desert, Anapra, Mexico is now the home of many of Mexico's poor. Without major utilities and services, many residents are forced to extreme measures to provide for themselves and their families. The east side of Anapra however, exhibits a stark contrast to the poverty of the west side. Here we find many of the regions elite with large houses and expensive luxuries. Many of the west side residents commute to the east for work. This geography of wealth illustrates the importance of relative location. The wealthy areas have connections to maquiladoras, which have the advantageous relative location along the United States border and close access to labor. While some of Mexico's poor choose to work in maquiladoras, others have to take a more desperate measure such a crossing the United States border. These crossers face endless dangerous obstacles in an attempt to leave their developing region for work in the developed region of North America just across the border. While many Mexican workers in the United States have legal permits, hundreds cross illegally when poverty leaves them no time to spare.

Operation Hold the Line

The regions of the United States and Mexico are considered formal regions since there is a concrete border between the two places. Cultural regions are more difficult to define however since they are subjective in nature with ideas flowing easily between one region and another. Cultural regions therefore, can cross and overlap formal regions. The El Paso/Juarez is an example of a culture region with El Paso in the United States and Juarez just across the border in Mexico. This "informal" region is called the borderland. At a local scale, the political border is an obstacle for immigration and work. At a global scale, the problem becomes more politically complex. "Operation Hold the Line" is a blockage strategy designed to deter illegal immigrants into the United States. While its goal was to provide a well-defined and peaceful border, many argue its efficacy since agent job satisfaction is low and many people still cross above or below the line in more dangerous environments. This has encouraged many Mexican workers to bring their whole families and stay in the United States permanently rather than attempting to cross back and forth many times.

DISCUSSION OF UNIT THEMES

Globalization is the increasing level of interconnections among people throughout the world. The speed and intensity of globalization increased considerably in the 1990s. Unit 1 in the Power of Place exemplified some of the forces that draw together or pull away places from each other and the implications of such processes. Whether it is the cultural landscape change in Vancouver, the ethnic mosaic of Strasbourg, the devolving Europe, civil unrest, or industrial location, all are examples of the importance of interconnectivity. In addition, the implications of these links vary tremendously according to whether the scale is global, regional, national, or local.

The type of region is also important. Regions can be formal or informal and can vary from political, law and order, economic to cultural, and environmental. Formal regions are clearly defined and objective, while informal regions are subjective and abstract. A clearly defined formal political region may include a variety of cultural informal regions and vice-versa. This overlap may lead to conflict such as the case of the borderland region. The importance of location is also exhibited in the borderland. Absolute location is the exact point on the earth's surface, or the exact latitude and longitude. Relative location is where something is relative to something else. The relative location of Mexico just south of the United States plays a key role in the conflict.

Above all else, it is the spatial perspective that sets geography apart as a discipline. This spatial analysis highlights the relationships among places resulting from linkages over the space of the Earth's surface. As seen in Program 1, these linkages can be analyzed not only on the ground, but also from space. Regardless of scale, the global spatial relationships cannot be ignored.

TEST YOUR KNOWLEDGE

Multiple Choice and True/False Questions

1. Which of the following is a way in which economic development can be seen from space?

   a. by seeing lots of lights in a nighttime view
   b. by noticing more green space
   c. by the number of people living in the area
   d. by the amount of air pollution seen in the air

2. Which of the following can be considered a first world or developed nation?

   a. India
   b. Brazil
   c. Japan
   d. Thailand

3. With approximately 5% of the world's population, which country uses 25% of the world's electricity?

   a. Canada
   b. United States
   c. Japan
   d. China

4. Analyzing the world in terms of relationships, patterns, and process is known as the __________ perspective.

   a. spatial
   b. historical
   c. geological
   d. none of the above

5. Which of the following is a key theme of globalization?

   a. expansion
   b. sovereignty
   c. separation
   d. interconnectivity

6. Relative location can be described as:

   a. the location of something relative to something else.
   b. the exact point on the earth's surface.
   c. the latitude and longitude of a point.
   d. both b and c

7. An example of a developing nation is:

   a. United States.
   b. Japan.
   c. Canada.
   d. Mexico.

8. Which of the following in an example of a formal region?

   a. the borderland
   b. Canada
   c. the Bible belt
   d. the cotton belt

9. True/False — “Operation Hold the Line” has proved very successful in dealing with border disputes and immigration policy.

10. True/False — Images from space can tell us much about the physical properties of earth, but nothing about humans.

Short Answer and Essay Questions

1. What can satellite imagery tell us about the world?
2. What makes the spatial perspective unique?
3. List examples of globalization.
4. How can relative location be advantageous or disadvantageous to a regional economy?
5. Discuss the importance of region in terms of immigration policies.

Map Exercises

1. Looking at the blank outline map, what variables were used to draw the formal state region boundaries?
2. Using the blank outline map, draw the borders for as many informal regions you can think of in the United States. How did you decide where to draw the lines?

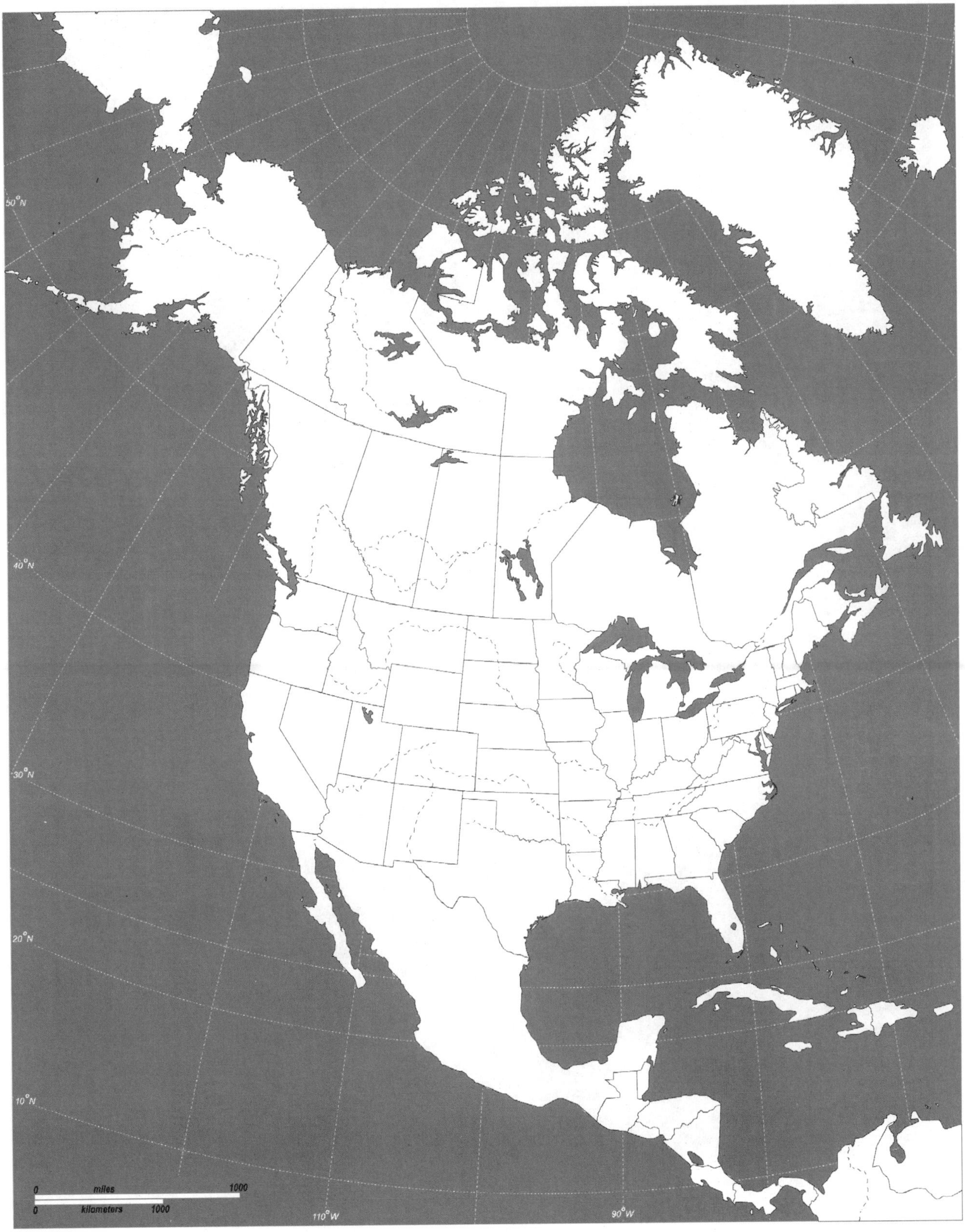
50° N
40° N
30° N
20° N
10° N
0 miles 1000
0 kilometers 1000
110° W
90° W

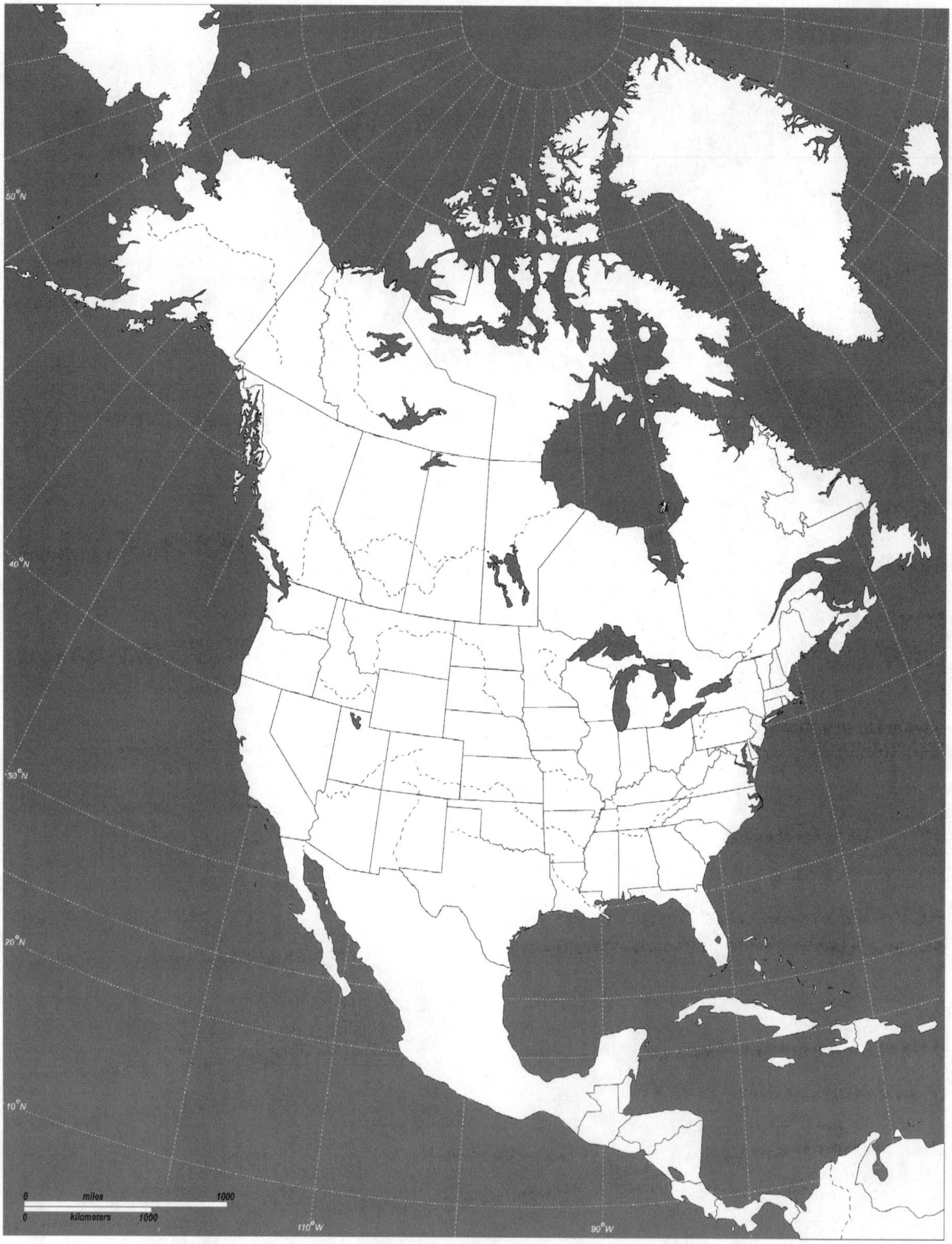
50°N
40°N
30°N
20°N
10°N
0 miles 1000
0 kilometers 1000
110°W
90°W

50°N
40°N
30°N
20°N
10°N
0 miles 1000
0 kilometers 1000
110°W
90°W

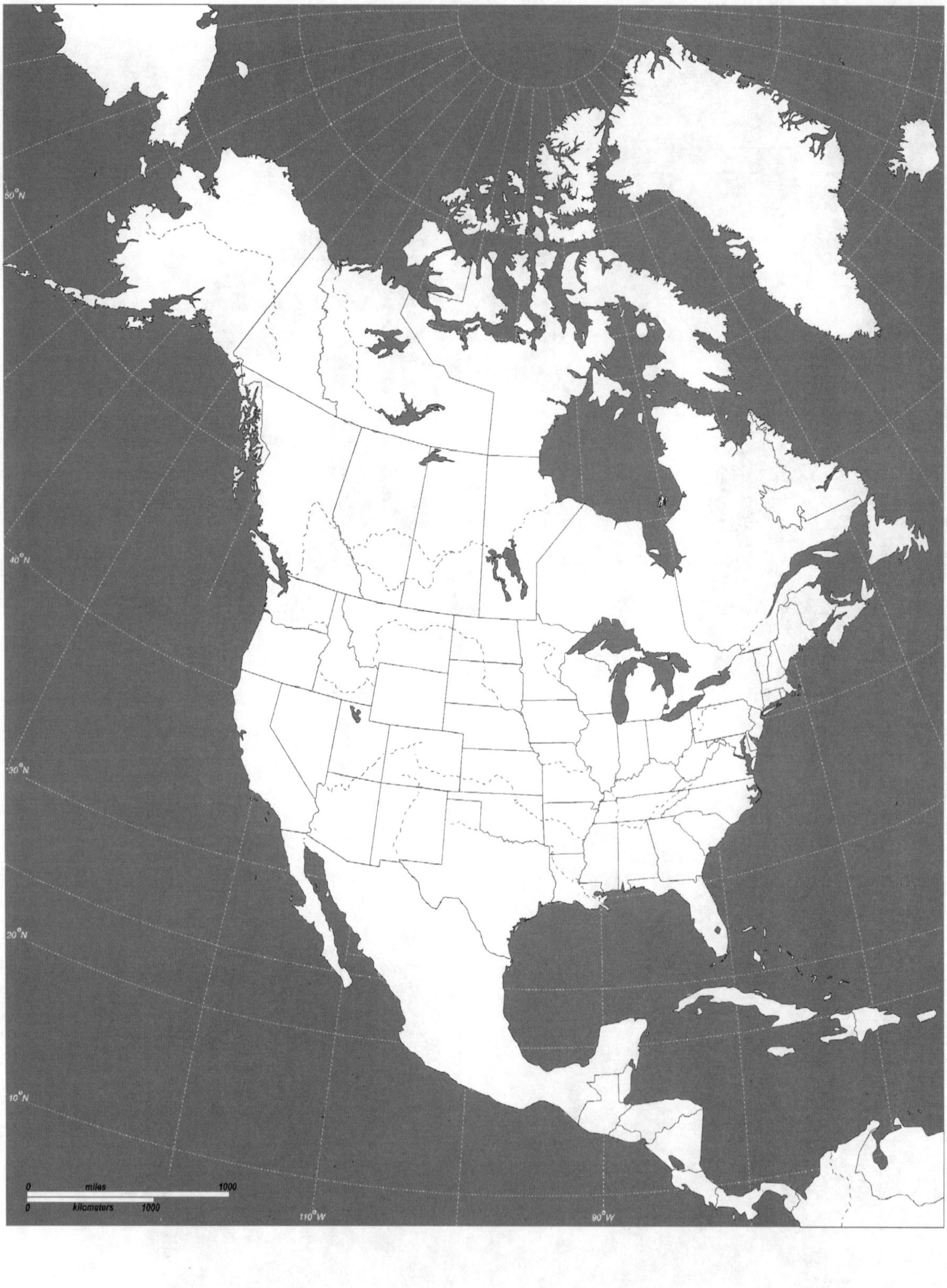

50°N
40°N
30°N
20°N
10°N
0 miles 1000
0 kilometers 1000
110°W
90°W

UNIT TWO: EUROPE

Unit Objectives

At the completion of the unit, you should be able to:

- understand Europe's cultural diversity.
- understand Europe's physical geography.
- understand the processes of unification, supranationalism, and balkanization.
- understand the processes of European past and present industrialization and urbanization.

PROGRAM 3

SUPRANATIONALISM AND DEVOLUTION

Key Themes

- **Cultural unification**
- **Balkanization**
- **Democratic diffusion**
- **Landscape preservation**
- **Locating capitals**
- **Core/periphery relationships**
- **Urbanization and urban renewal**

Strasbourg: Symbol of a United Europe

Strasbourg is a European cultural crossroads. Remnants of its German and French past are evident in the cultural landscape. Strasbourg is an excellent example of a nation forging toward ethnic unity. Both cultures exist side by side. Germans can freely cross into Strasbourg without stopping at customs and dozens of French also travel and work in Germany. Strasbourg's importance grew in 1949 when the Council of Europe located there. European communities soon followed. In the past, Strasbourg was caught in between conflicts between Germany and France. Today however, Strasbourg serves as the future image of European cultural and ethnic unity.

Slovakia: New Sovereignty

While Strasbourg is a good example of European cultural unity, Slovakia is an example of Europe splitting apart. Czechoslovakia peacefully separated from the Czech Republic and Slovakia in 1993. Although many Slovakians felt discrimination by the Czech, not all people agree that forming an independent state was the best decision. Ethnic groups such as Hungarians and Romani, left on the Slovakian side, are not ethnic minorities and are wary of the transition. The Czech's excellent relative location to Western Europe will no doubt prove beneficial. Without the power of the Czechs however, the Slovakian economy may languish.

PROGRAM 4

EAST LOOKS WEST

Key Themes

- **Locating capitals**
- **Looking beyond borders**
- **Political diffusion**

Berlin: United We Stand

During the Cold War, Germany existed as two separate parts: West Germany and East Germany. To complicate matters, Germany's capital, Berlin, was also divided into east and west by the Berlin Wall, however the entire city existed within the greater communist region of East Germany. The wall came down in 1989 and the city was united. True economic and cultural unity however did not instantly appear. East Berlin economically languished behind its western capitalist counterpart and it would take more than removing the wall to truly unite as one city. Economic endeavors have been quite successful, as well as urban renewal projects. The most important development however, has been to put the capital of the new Germany in its original location of Berlin. Berlin's relative location, although central within Germany, links it with both Western and Eastern Europe. Its redevelopment showcases the new economic and political might of a united country.

Poland: Diffusion of Democracy

After the fall of the iron curtain, one of Poland's main objectives was to spread democracy throughout the country. The challenge for Poland is its relatively large rural population which is often more resistant to change than urban populations. After the initial years of capitalism did not prove as successful as expected, it became evident that the basic skills of discourse, compromise, and civic engagement have not yet been mastered. Democracy thus far has been popular in urban centers, but resistance is strong in urban areas due to isolation and unemployment. A team of young Poles developed organization strategies such as Ping-Pong tournaments. While these may seem trivial, the goal was to slowly engage the rural populations to become active participants in their community. Poland's geographic realities play a key role and the context of place is very important when choosing new communities for their projects.

PROGRAM 5

THE TRANSFORMING INDUSTRIAL HEARTLAND

Key Themes

- **Urbanization**
- **Urban development**
- **Urban renewal**

Liverpool: A Tale of Two Cities

Once a major industrial city in Europe, Liverpool, like many industrial and manufacturing cities began to languish in the age of the rise of the service sector. Liverpool claimed to be one of Europe's most important harbors for nearly 50 years until things started to change in the middle of the 20th century. The economic strength of Europe beyond England was growing, and Liverpool was on the periphery. Many docks closed and unemployment was at all-time highs. Like many cities facing similar problems, Liverpool opted to redevelop the riverfront through several urban renewal projects. Many involved gentrifying old port buildings into office space and upscale loft apartments. Businesses and waterfront hotels soon followed. The European Union has proved to be a key element in Liverpool's success. The EU was most concerned with poverty stricken areas and EU funds have aided in several projects including museums and universities. Liverpool is now attracting high tech, cultural, and service industries, which have an international market and almost guarantee Liverpool a brighter future.

Randstad: Preserving the Green Heart

The Randstad is a horseshoe shaped continuous urban zone around an undeveloped green zone that is marshy and difficult to develop. There are seven million people living in this area and connectivity between places is extremely important. In the early 1990s, this need for connectivity sparked a project to build a high-speed rail line through the heart of the green zone, shortening the commute time between cities. This rail line however would pass through one of the areas last green spaces and therefore would have severe environmental impacts. It was decided that the rail line would include a tunnel that would not interfere with the most important area of the Green Heart.

PROGRAM 6

CHALLENGES IN THE HINTERLANDS

Key Themes

- **Periphery**
- **Harsh environment**
- **Human/environment interaction**
- **Resource management**

Andalucia: Developments in the Hinterlands

Situated in a peripheral region of Spain, Andalucia is known for its tourism, folklore, and tradition. In the 1970s, this region obtained relative autonomy because of its old economic and political traditions. Since its autonomy, thousands of people relocated to the area with the goal of slowly becoming a core area by adding high-tech industry to its traditionally rural economy. Unfortunately, new peripheries have emerged with cheaper labor and Andalucia is losing its attractiveness.

Iceland: Edge of the Habitable World

Fish are the main export of Iceland and contribute to more than 70% of its economy. While most nations find EU membership advantageous, Iceland has resisted since it would result in opening their fishing grounds to continental boats. Iceland's strategic location mixes cold water with warm water which pulls nutrients from the ocean floor providing a wealth of food for fish. Iceland's location over an ocean hotspot provides geothermal power and its waters are tapped for hydroelectric power, making electricity on Iceland extremely cheap. These geographic advantages can be dangerous as well. The mixing of waters makes ocean currents rough and treacherous and Iceland is always susceptible to volcanic activity. Their main threat however, is a decrease in their cod population. Over-fishing has led to a drop in numbers. Smaller fleets are now encouraged to protect this valuable resource.

DISCUSSION OF UNIT THEMES

Europe contains a rich variety of physiography, economy, and culture. Each place has its own unique way of dealing with their specific problems. Supranationalism has played a key role in the success of several countries and Europe has made substantial progress toward international economic collarboration. Specific cities such as Strasbourg, are celebrating their ethnic differences. Newly united cities such as Berlin and new emerging democracies such as Poland are slowly but surely making their mark on the world economy. In other cases, such as Slokavia, peaceful relations resulted through separation rather than unification.

Much of Europe's population is urban and urbanization continues as Europe moves through the height of industrialization, the domination of manufacturing, to the rise of the service sector. Urbanization however, comes with challenges. These challenges could be in the form of development versus the environment as we saw in Randstad, or urban renewal and gentrification as we saw in Liverpool.

Places on the economic periphery need to work extra hard to become a major player. Places such as Andalucia struggle to become part of the core, while places such Iceland take advantage of their peripheral physiographic advantages.

TEST YOUR KNOWLEDGE

Multiple Choice and True/False Questions

1. The Russian Communist Party argued for democratic centralism. Which of the following best describes the beliefs of democratic centralism?

   a. The Communist Party, the political party of the working class, was the only true representative of the people, and therefore, the only party with the right to govern.
   b. All political parties are created equal and thus deserve equal opportunity to rule the country.
   c. The Communist Party represents the wealthy elite of Russia, for it is the rich who have the most power.
   d. There should be no central government and the people of each Russian state should govern themselves.

2. Which of the following best describes the European climate?

   a. All of Europe falls under a Mediterranean climate with long, dry summers and mild winters.
   b. Most of Europe experiences midlatitude west coast climates.
   c. Much of Europe is tropical with very hot and humid summers.
   d. Only the eastern portion of Europe receives more than 10 inches of rain per year.

3. Which of the following rivers is not found in Europe?

   a. the Danube
   b. the Rhine
   c. the Seine
   d. the Yangtze

4. After World War II, which three countries joined together to form "Benelux"?

   a. Belgium, Norway, Luxembourg
   b. Belgium, the Netherlands, Luxembourg
   c. Berlin, the Netherlands, Liechtenstein
   d. Bosnia, Nice, Luxembourg

5. Post-war construction included renovation of poor and dilapidated housing, which increased the value of housing and raised neighborhoods from low-income to higher-income. This process is known as:

   a. gentrification.
   b. assimilation.
   c. agglomeration.
   d. blockbusting.

6. Which of the following was not part of the former Yugoslavia?

   a. Albania
   b. Montenegro
   c. Croatia
   d. Slovenia

7. Which of the following statements regarding Slovakian independence is false?

   a. Slovakians languished because the Czech side was dependant on them.
   b. Many Slovaks were upset with the lack of equality.
   c. Border follows approximately the historic limitations of the ancient Czech and Slovak federal republics.
   d. The decision to divide was primarily a political decision and not all Slovaks agreed that it was beneficial.

8. After a reunified Germany, the decision was made to put the capital in its original historic location:

   a. Bonn
   b. Berlin
   c. Frankfurt
   d. Munich

9. Which of the following is NOT an example of supranationalism?

   a. European Union
   b. Balkanization
   c. Benelux
   d. United Nations

10. The most northern capital in the world is:

   a. London.
   b. Berlin.
   c. Reykjavik.
   d. Oslo.

11. True/False — Resistance to democracy in Poland is most evident in the small towns and villages.

12. True/False — The people of Slovakia voted unanimously for sovereignty.

Short Answer and Essay Questions

1. What are some of the ways in which Liverpool is attracting new business and investment? How has the EU been beneficial?
2. How has Strasbourg been successful with cultural unity?
3. How has the physical geography of Iceland contributed to energy production?
4. Describe the main challenges Poland faces with the diffusion of democracy?
5. What are the main argument surrounding the construction of a railway through the Randstad region of the Netherlands?

Map Exercises

1. Place the following physical features on the outline map: the Alps, Mediterranean Sea, the Seine river, and the Po Plain.

2. Using the blank outline map, locate the following European countries: Poland, Iceland, Spain, England, the Netherlands, France, and Slovakia.

3. Using the blank outline map, make a dot for the location of the following European cities: Strasbourg, Liverpool, and Reyjavik.

70°N
0°E/W
60°N
50°N
40°N
0 miles 500
0 kilometers 500

70°N
0° E/W
0
miles
500
0
kilometers
500

70°N
0°E/W
60°N
50°N
40°N
0
miles
500
0
kilometers
500

70°N
0°E/W
60°N
50°N
40°N
0 miles 500
0 kilometers 500

70°N
0°E/W
60°N
50°N
40°N
0 miles 500
0 kilometers 500

UNIT THREE: RUSSIA AND NEIGHBORING COUNTRIES

Unit Objectives

At the completion of the unit, you should be able to:

- understand the factors involved in a resource-based economy.
- appreciate the problems faced by transitioning from a collective to market system.
- understand the importance of location.
- understand the influence of topography on ethnic diversity.
- understand the role of nationalism in civil unrest.

PROGRAM 7

NORTHWEST CONTRAST

Key Themes

- **Privatization**
- **Aging population**
- **Environmental constraints**

St. Petersburg: Russia's Window on the West

The fourth largest city in Europe, St. Petersburg has a unique northern location on the Neva River delta. Its location provided for a close relative location to the rest of Europe, plus access to an outside port. This combination helped St. Petersburg become Russia's cultural capital until the Russian revolution. During the Soviet era, its name was changed to Leningrad and the capital moved to Moscow. Since the fall of the Soviet era in 1989 however, St. Petersburg regained its original name and along with the rest of Russia, is in the process of a dramatic transition. One of the greatest transitions is in housing. Once entirely owned and managed by the state, housing now can either be state or privately owned. Unfortunately privatization has increased housing costs and the increase in incomes has not grown significantly. This has caused an economic gap where the new homeowners are those in the wealthy trade sector dubbed the "new Russians." Many others now fear a future no longer guaranteed by the government. While the St. Petersburg location is still ideal, many wonder about the future of Russia in a free market economy.

Vologda: Russian Farming in Flux

Once in the hands of the state, most of the collective farms have been privatized after the communist era. These new enterprises are still quite large and in many ways managed the same way, which means only a few people can start small family farms. The city of Vologda lies in the northern part of Russia where soils are not very fertile. The physical environment of the area is more suited to dairy farming. Dairy farms are now trying to transition to market conditions in which housing and schooling for farm workers is now private instead of arranged by the state. While this may make farmers more efficient, money is scarce. Rural populations are dwindling with rural to urban migration and an aging population. Short summers make for little work for most of the year. The regions abundant forest resources may help with the local economy, but agricultural and forestry still have an uncertain future.

PROGRAM 8

HOLDING THE HINTERLANDS

Key Themes

- **Ethnic relations**
- **Resource exploitation**

Dagestan: Caucuses Disconnect?

Dagestan lies on Russia's southern border, neighboring the volatile region of Chechnya. Chechnya conflict arises out of nationalism and religious differences with Russia. Dagestan has a unique geography where higher elevation brings moisture to the area, making it one of few places where people can live at high altitudes. The mountainous topography has led to small clusters of culturally distinct villages. This diversity has reduced nationalism. This coupled with differing experiences of Soviet rule explains the peacefulness of Dagestan as compared to Chechnya. Most residents also acknowledge that due to poor resources and extreme environment, it is unlikely that they can survive as an independent state.

Bratsk: The Legacy of Central Planning

Bratsk lies in the heart of the inhospitable climate of Siberia. Despite freezing temperatures, approximately 300,000 people live here. The city of Bratsk was built along with construction of the largest hydroelectric plant in world on the Angara river. Many argue that the dam was not a necessity, but built as a national symbol of the success of communism. Factories were soon built that took advantage of the plentiful electricity. Bratsk became known as the first Territorial Production Complex (TPK), which is a complex of mutually related factories that utilize the natural and economic resources of the region. Environmental pollution is now a major concern, as well as privatization. Fuel costs may rise making living conditions too expensive. Still many choose to stay in Bratsk, despite the uncertain future.

DISCUSSION OF UNIT THEMES

Many of the case studies of this unit discuss Russia's transition to privatization. Under the Soviet government, the state owned everything, including natural resources. Factory managers paid only minimal costs for the fuels they used, giving them little incentive to use fuels efficiently or find cheaper fuels. Fuel producers, also state agencies, lacking competition or having to worry about paying for costs of production, saw no need to develop new types of fuels. As fuel sources ran out in the western Soviet Union, new ones were tapped in the north and in Siberia. Exploiting these locations and transporting the fuels cost a lot of money and the entire economy carried the burden (*Contemporary World Regional Geography*, Bradshaw/White/Dymond, © 2004, McGraw-Hill Education). Examples of this exploitation were seen in Vologda and Bratsk. Farms in the remote and rural of Vologda still operate in a collective Soviet manner. Its extreme northern location brings cold temperatures and poor soils discouraging new growth. The exploitation of Bratsk has proved a bit more successful and it is only its resources that have allowed this city to survive in the bleakest of physical conditions.

Dagestan exemplifies the connection between topography and ethnic identity. Dagestan is located in a very mountainous and cold region of Russia. This, along with its location at the northern edge of the Islamic world, has contributed to it becoming an ethnically diverse region. Russian is spoken to bridge the diversity of language and also links Dagestan to Russia. Ties to Russia have proved economically advantageous to Dagestan and affords them a more peaceful fate than their neighbors in Chechnya.

TEST YOUR KNOWLEDGE

Multiple Choice and True/False Questions

1. Which of the following is not a spatial advantage of St. Petersburg?

    a. landlocked plateau location
    b. close proximity to Europe
    c. on a river delta
    d. port access

2. Which of the following best describes a Territorial Production Complex or TPK?

    a. a foreign commercial enterprise in a remote oversees location
    b. a complex of factories, which are dependent on imported resources
    c. an alliance of companies evenly distributed throughout the entire country
    d. a complex of mutually related factories in a region that together utilize the natural and economic resources of the region

3. The success of Bratsk was largely dependant upon which of the following?

    a. the great physical environment
    b. the building of a large hydroelectric plant
    c. the fall of the Soviet era
    d. new trade agreement with the United States

4. The world's deepest lake is found in:

    a. Central Russia.
    b. Siberia.
    c. Ukraine.
    d. none of the above

5. Conflict in Chechnya is mostly due to:

    a. poverty.
    b. nationalism.
    c. religion.
    d. b & c

6. Which of the following is not a characteristic of Dagestan?

   a. high altitude
   b. plentiful rainfall
   c. ethnically homogenous
   d. majority Muslim

7. The current capital of Russia is:

   a. Leningrad.
   b. Moscow.
   c. St. Petersburg.
   d. Vologda.

8. True/False St. Petersburg's housing market has become affordable to all since privatization.

9. True/False Vologda's lush climate makes it an excellent region for agriculture.

10. True/False Dairy farming in Vologda has been completely unsuccessful since privatization.

Short Answer and Essay Questions

1. Why did the Soviet Union decide to build the city of Bratsk in the 1950s?
2. What are some variables that might explain the differences between Dagestan and Chechnya?
3. Has privatization helped Russian agriculture? Why or why not?
4. Describe the main challenges for St. Petersburg in a post-Soviet society.
5. Describe the spatial locational advantages of St. Petersburg.

Map Exercises

1. Using the blank outline map, find the following locations: St. Petersburg, Siberia, Vologda, Bratsk.

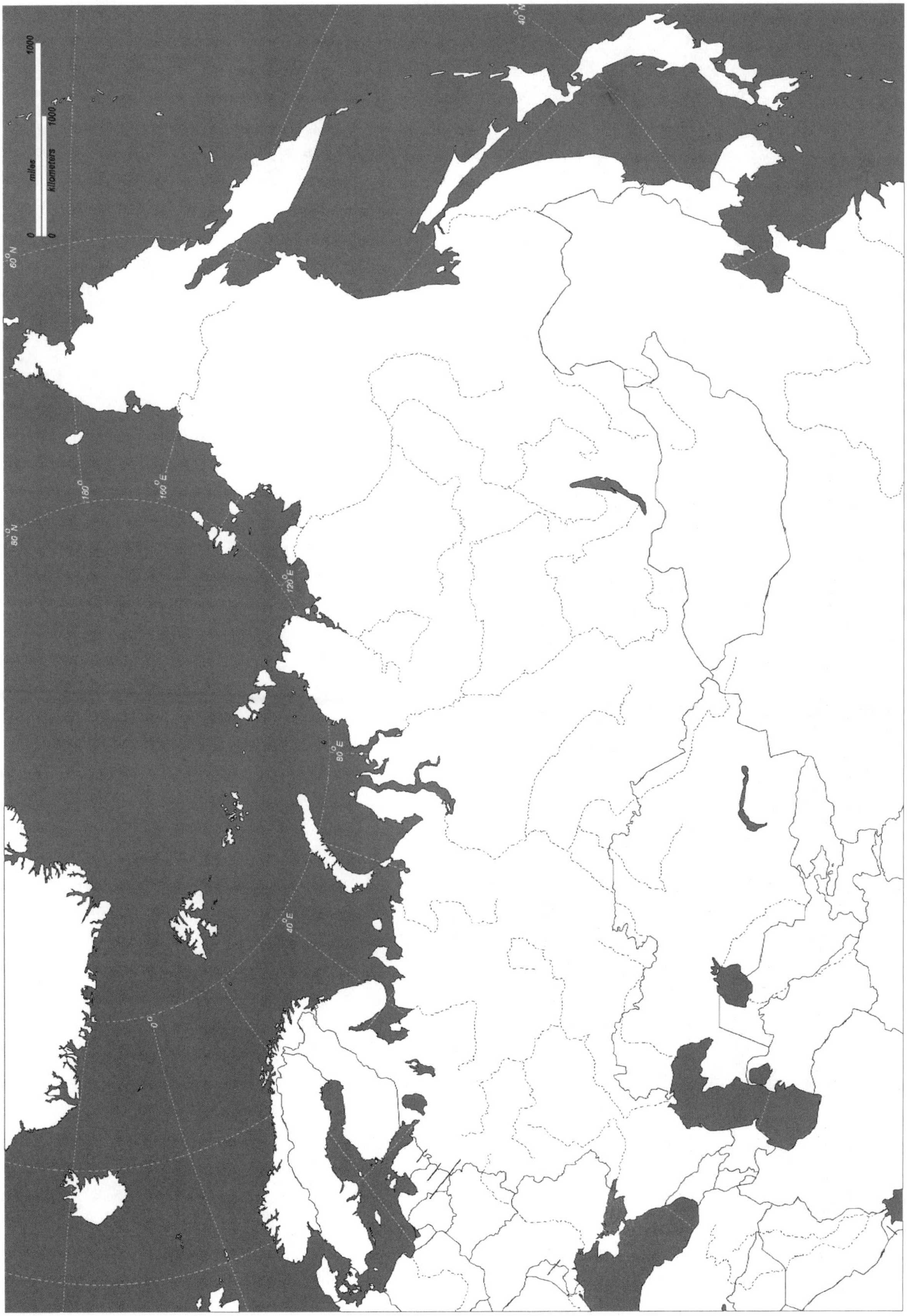
0 miles 1000
0 kilometers 1000
60° N
80° N
40° N
180°
150° E
120° E
80° E
40° E
0°

0
miles
1000
0
kilometers
1000
60° N
80° N
40° N
180°
150° E
120° E
80° E
40° E
0°

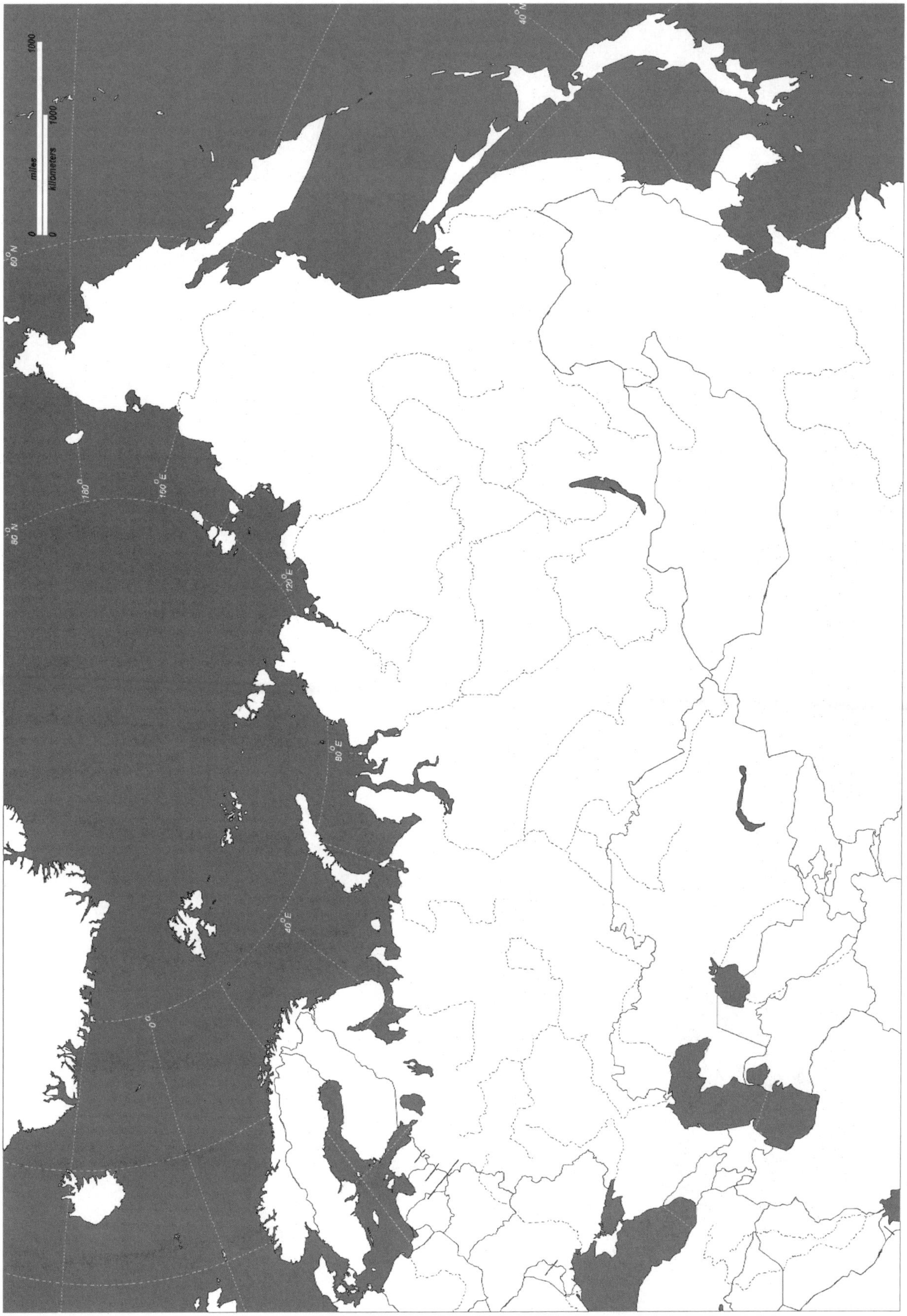

0 miles 1000
0 kilometers 1000
60° N
80° N
40° N
180°
160° E
120° E
80° E
40° E
0°

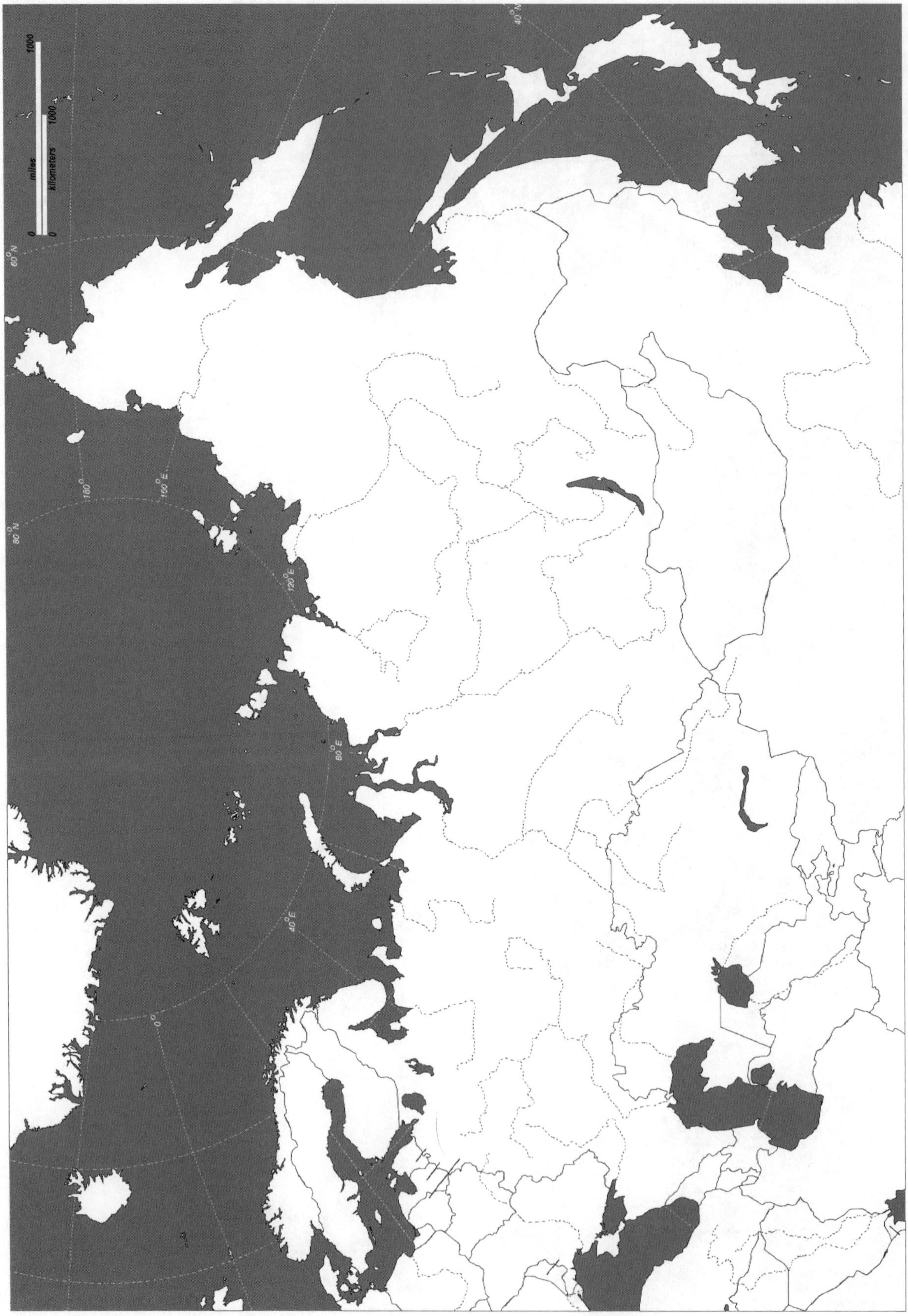
0 miles 1000
0 kilometers 1000
40°N
60°N
80°N
180°
160°E
120°E
80°E
40°E
0°

UNIT FOUR: EAST ASIA

Unit Objectives

At the completion of the unit, you should be able to:

- understand the process of urbanization and suburbanization.
- understand the impact of globalization on local cultures and rural areas.
- understand what factors contribute to industrial development.
- understand the influence of culture on development.
- understand the impact of urbanization on agriculture and housing.

PROGRAM 9

CHANGES ON THE CHIANG JIANG

Key Themes

- **Industrialization**
- **Rural enterprises**

Shanghai: Head of the Dragon

Shanghai can be considered China's most important city. In recent decades it has experienced unprecedented urban and economic growth as it strives to enter the ranks of the world's greatest megacities. Shanghai is strategically located on the eastern coast of China at the delta of the Chang Jiang—China's longest and most important river. Shanghai has long been globally influenced. Many countries including Britain, America, and France had trading settlements here. Their influence can still be seen in the urban landscape. Shanghai's importance began to decline after invasion of the Japanese and later Communism and the city began to languish. In the 1990s however, this changed. Lifted restrictions on development and foreign investments have helped Shanghai once again gain power and it is now slated to be the second busiest port after Hong Kong. Shanghai is taking steps to ensure its position, such as building a deeper port to accommodate larger ships and streamlining the customs process. Increased development has led to the common problems associated with densely populated, congested cities: lack of services, traffic, air pollution, and inadequate housing. In an effort to combat these problems and make room for increased growth, the Pudong area was developed in 1990. Pudong is located across the Huang river from old Shanghai and offers acres of land for industrial development. Pudong's tall buildings and glass skyscrapers pose a dramatic contrast to the colonially influenced old Shanghai and is a symbol of China's new position in the 21st century.

Sijia: Small Town, Big Change

Sijia is a small, agricultural town located in interior China. This small rural town is also the location of a garment factory. Industrial growth, particularly factories, is becoming more common in rural China and has changed rural life. Within China there are major contrasts between the wealthier and richer provinces, the rural and urban, and the agricultural and industrial sectors. Sijia lies at the crossroads of these contrasts since it is located along the Chang Jiang river and only 200 miles from the coast. Industrial expansion in rural China has mostly taken the form of small, locally managed companies such as the garment factory in Sijia. These "township enterprises" operate differently than national corporations and are beginning to attract foreign investment. Most residents of Sijia are employed by the factory and standards of living have greatly improved. Most residents now have access to modern technologies such as televisions, cell phones, DVD players, and the internet. Increased labor needs in rural areas may prompt the government to change current policies regarding rural to urban migration.

PROGRAM 10

THE BOOMING MARITIME EDGE

Key Themes

- **Globalization**
- **Economic tiger**
- **Newly developing country**

Guangdong: Globalization in the Pearl

Globalization has had a profound impact on the global economy. The "global production system" is the heart of this process and allows manufacturers to produce goods anywhere on the globe. Guangdong province is home to many global factories, including Nike, and many Chinese have migrated there for work. Guangdong's proximity to Hong Kong and its cultural links to the rest of the region, makes it an ideal location for foreign investment. Cheap labor is the driving force behind the global production system. Companies such as Nike can pay workers a relatively small salary in comparison to a U.S. equivalent; however, compared to a typical Chinese salary, it is quite high. Over half a million factories are now located in Guangdong. Advancements such as the freight container and computer tracking have significantly improved efficiency of the global system and raw materials can be easily and inexpensively shipped from other parts of the world to the factories. Globalization is changing the face of the world and its impact on the local environment is quite dramatic.

Taiwan: High-Tech Tiger

A few decades ago, Taiwan was largely an agricultural society. Today Taiwan is one of the top producers of high tech computer and telecom products and developing at rapid pace. In the 1970s many Taiwanese went to study in the United States to learn about technology. These Silicon Valley returnees now work in companies located in Hsinchu Science Park, a government funded park designed to increase the high tech industry in Taiwan. Government incentives and access to good universities made locating in the science park very attractive. Hsinchu has been a huge success and has created such a demand that other similar parks are being developed in other areas of the country. Taiwan's ties to both the Silicon Valley and China make it a key player in the region.

PROGRAM 11

A CHALLENGE FOR TWO OLD CITIES

Key Themes

- **Industrial development**
- **Urbanization**
- **Central planning**
- **Cultural mosaic**

Lanzhou: Confluence of Cultures

Lanzhou, located in the rural western portion of China, has a population of 2.6 million with a diverse ethnic composition of both Han and Muslim. Lanzhou sits below the Huang He river and lies along the famous Silk Road. As China strives to develop its western sector, Lanzhou shows great economic promise. Lanzhou also sits at a cultural crossroads linking the Han of the east to the Muslims of the west. The Hui are the largest Muslim group and consider themselves ethnically distantly separate from the Han. Although predominantly Muslim, there appears to be no conflict between Muslims and Han Chinese in Lanzhou. Tensions only revolve around government preferential treatment that is given to rural Hui and not rural Han. Given recent events, the Islamic western portion of China, including Lanzhou, may receive more attention.

Shenyang: Hope for China's Rust Belt?

Over 1000 factories are located in Shenyang, China's fifth largest city. The aging factories are struggling to compete with the growing economy of the south. Old equipment and inefficiency are keeping them behind and quickly outdating them. These large state operated factories face the problem of how to acquire new equipment and more efficient methods. China is slowly trying to upgrade equipment and transition from an in-house production system to more specialized manufacturing. While foreign investment is increasing in the region, it is still far behind southern China.

PROGRAM 12

SMALL FARMS, BIG CITIES

Key Themes

- **Urban growth**
- **Agriculture in adverse climates**

Northern Japan: Protecting the Harvest

Rice is Japan's stable food and urban growth in the south is forcing Japan to grow rice in areas not conducive to agriculture. Rice farmers in northern Japan must contend with "yamase," a cold wind that blows through the region from June to August. This along with temperature and available light significantly shorten the growing season. Hybrid seeds that can grow in this environment have been developed. Farmers are forced to constantly check their paddies during the roughest time of the year. The northeast has been slower to develop than other parts of Japan, making it a logical choice for agriculture despite the climatic conditions. With government investment, Japan has turned parts of this region into prime rice producing areas.

Tokyo: Anatomy of a Mega-City

Japan's mountainous topography has rendered it largely uninhabitable. This physical geography has led to a concentration of people where there is flatter terrain and has made Tokyo the most densely populated place on earth. As a result, living costs have skyrocketed and many Japanese are moving to the suburbs in search of affordable housing. Japan's efficient transportation system has allowed for connectivity between the suburbs and cities. Suburbanization has led to longer and longer commute times, and a 90 minute commute from home to work is not uncommon. Small edge cities are developing to ease the over-crowding in Tokyo.

DISCUSSION OF UNIT THEMES

Two powerful regions, China and Japan dominate East Asia. In recent decades, China has strived to once again hold a powerful place in the world economy. Shanghai is experiencing a wave of rapid economic development. In the late 1990s, the new economic zone in Pudong across the Huang Pu River from Shanghai already had 4,000 enterprises and 400,000 workers. Even towns in the interior of China, such as Sijia, are experiencing rapid growth (*Contemporary World Regional Geography*, Bradshaw/White/Dymond, © 2004, McGraw-Hill Education).

Globalization has been at the forefront of this economic growth. Factories dot the urban and rural landscape as foreign investors look for ideal locations and cheap labor. This has had a dramatic impact on the rural landscape. From the 1960s to mid-1970s, the rural-urban basis of the registration system established in the 1950s, prevented urban-rural people from becoming legal residents of cities. From the late 1970s, new industries recruited labor in the countryside on temporary contracts; such workers now make up to 10% of urban inhabitants. Places such as Guangdong and Taiwan are at the forefront of economic development (*Contemporary World Regional Geography*, Bradshaw/White/Dymond, © 2004, McGraw-Hill Education).

Many regions of China are now in the transition from "old" to "new." "Collective" systems are being replaced by more efficient free market systems. Peripheral regions must not only adjust to new market cities, but also cultural differences as well. The Han Chinese make up 92% of the Chinese population. They are of similar appearance to most of the national minorities but derive their ethnicity from the emergence of the Chinese administrative culture in the A.D. 200s and 300s. Most minority groups live along the northern borders with Russia and Mongolia and in western China. Lanzhou, in western China, must balance the power between the Han and Muslim Chinese (*Contemporary World Regional Geography*, Bradshaw/White/Dymond, © 2004, McGraw-Hill Education).

After the mid-1970s, improving transportation encouraged people to move out of central cities of Japan that were being stifled by congestion and pollution. This was more a process of suburbanization, when people move to the outskirts of existing cities, than of counterurbanization, when they move away from metropolitan centers to small towns in rural settings. The expansion of metropolitan areas has led to agricultural problems. Japanese farms occupy small areas of suitable land within the hilly and mountainous country, often intermingled with housing and industry. Urban and industrial users also compete for land, paying higher prices than farmers could afford without subsidy. Farmers must now battle harsh weather in the northern regions of Japan in order to maintain their positions as rice producers (*Contemporary World Regional Geography*, Bradshaw/White/Dymond, © 2004, McGraw-Hill Education).

TEST YOUR KNOWLEDGE

Multiple Choice and True/False Questions

1. The longest river in China is the:

    a. Ganges.
    b. Chang Jiang.
    c. Indus.
    d. Tibetan.

2. The Pudong area of Shanghai was built primarily to:

    a. accommodate new growth.
    b. house new immigrants.
    c. provide housing for dam workers.
    d. be a communist island within a new democratic society.

3. Factories run by a municipal township, village council, or private individual are called:

    a. special economic zones.
    b. national corporations.
    c. free zones.
    d. township enterprises.

4. Which of the following makes Guangdong province attractive to foreign investors?

    a. cheap labor
    b. proximity to Hong Kong
    c. cultural links to the region
    d. all of the above

5. One of Taiwan's advantages is long historical ties with:

    a. Australia.
    b. MIT.
    c. Silicon Valley.
    d. Hong Kong.

6. The world's most densely populated megalopolis is:

    a. Shanghai.
    b. Kyoto.
    c. Shenyang.
    d. Tokyo.

7. True/False China is ethnically homogenous, with Han Chinese being the only ethnicity.

8. True/False While globalization has had profound effect on the global economy, it currently has had little impact on local environments.

9. True/False China currently allows free rural to urban migration with no registration required.

10. True/False The Chang Jiang River is relatively shallow making it difficult for larger ships to transport goods into the interior of China.

Short Answer and Essay Questions

1. Describe which variables contribute to Tokyo being the world's most densely populated region?

2. Why doesn't Japan grow rice in a more suitable area?

3. What makes Guangdong attractive to foreign investors?

4. Are there any disadvantages to the rapid growth of Shanghai?

5. What are the impacts of globalization and industrial development on rural populations?

Map Exercises

1. Using the blank outline map, find the following locations: China, Japan, Chiang Jiang River, Huang He River, Shanghai, Tokyo, Taiwan, Taipei, Shenyang.

20° N
10° N
0° N/S

20° N
10° N
0° N/S

20° N
10° N
0° N/S

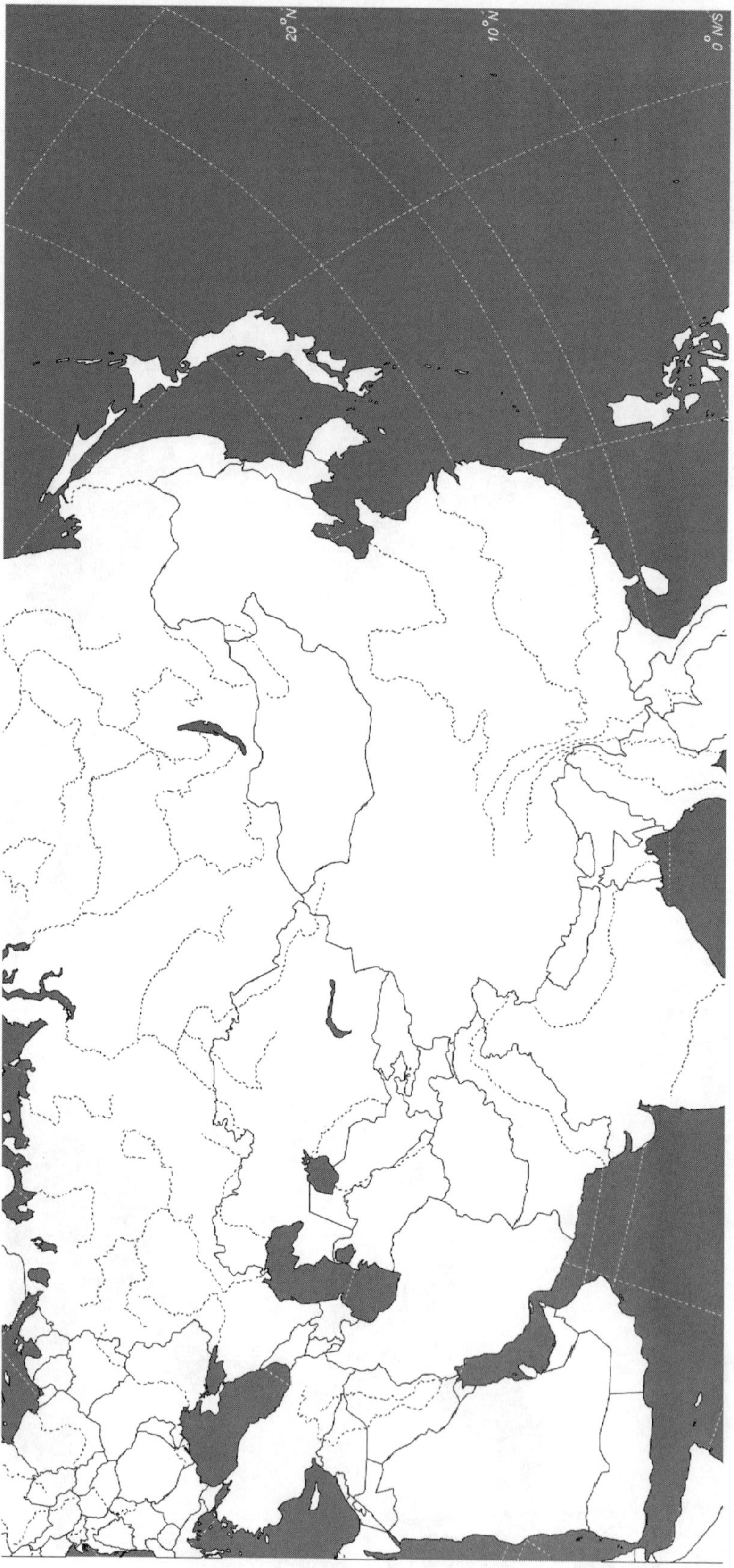
20° N
10° N
0° N/S

UNIT FIVE: SOUTHEAST ASIA AND SOUTH PACIFIC

Unit Objectives

At the completion of the unit, you should be able to:

- understand the challenges of a physically fragmented country.
- understand the relationship between natural resources and economic power.
- understand the importance of spatial location.
- understand the relationship between location and trade.

PROGRAM 13

THE MAINLAND

Key Themes

- **Natural resources**
- **Sustainable exports**

Laos: Isolated Heart

Laos is Southeast Asia's only landlocked country. Once isolated, the Friendship Bridge now connects Laos to Thailand. Laos is still a very poor country, despite its rich resources. One advantage to its mountainous topography, along with being surrounded by water, is its potential for hydroelectric power. Laos produces more electricity than it consumes and sells the surplus to neighboring countries. Laos has more than 60 potential dam sites and construction is build using the BOOT scheme (build, operate, own, transfer). The BOOT scheme uses foreign investment for construction and operation for a few decades and then the dam gets transferred back to the Laos government. This method is controversial however since local people rarely have a voice. Construction sites usually target minorities and hinder the subsistence farming of the local people. There are also environmental impacts that are often ignored. Dam construction does bring capital into Laos however and could help build a better infrastructure, providing better access to health care and education for much of the rural population. As Laos enters the global economy, this is a tradeoff that needs to be seriously considered.

Vietnam: Fertile Dreams

Vietnam is one of the world's largest rice producers due to the fertile Mekong Delta. In 1988, the collective system was dissolved and many mills are now privately owned, resulting in an increase in farmer productivity. Ho Chi Min City is the center of the rice industry. Its population has grown from 1 to 1.5 million in the last decade. Ho Chi Min City has an ideal coastal location and its history of colonial rule has left it with an excellent transportation system. This has allowed rice to not only feed their country, but become a sustainable export as well.

PROGRAM 14

THE MARITIME CONNECTION

Key Themes

- **Tourism**
- **Ethnic relations**
- **Ethnic mosaic**

Indonesia: Tourist Invasion

Bali, one of Indonesia's 13,000 islands, has been a global tourist destination for over 50 years. Like many countries that rely heavily on tourist dollars, tourism can be seen as a threat to local culture and maintaining a sense of identity is challenging. Bali is unique in that most residents are Hindu, while the rest of Indonesia is predominately Muslim. Seventy percent of the locals now rely on tourism as their main source of income. The Balinese believe however, that they can successfully balance tourism with culture and the natural environment.

Multicultural Malaysia

Malaysia is a multicultural country with approximately 60% of the population Malay, 30% Chinese, and 10% Indian. Malaysia has relatively little ethnic tension and ethnic groups have interacted successfully for decades. In 1969 there was ethnic violence after an election and social policies were soon implemented which curtailed the foreseen economic might of the Chinese. The government strived for economic equality for all ethnicities. This was done primarily by bringing rural residents, mainly Malay, into the Chinese dominant cities. Many Chinese left the country, but those who stayed enjoy a country that is experiencing great economic growth. Breaking the identification of ethnic group and occupation was very instrumental. Malays are no longer considered only farmers and the Chinese only merchants. Both ethnicities now work side by side. Should peaceful ethnic relations continue, Malaysia may very well become the next economic tiger.

PROGRAM 15

GLOBAL INTERACTION

Key Themes

- **Spatial location**
- **Trade**

Singapore: Gateway to Southeast Asia

Singapore is one of the wealthiest cities in the world. Singapore has a central location in the region and a good infrastructure, making it attractive to foreign companies. Singapore's port and airport are among the busiest in the world, although the city-state is only 400 square miles. Its relative location provides a direct connection to Asia, Europe, and the United States and its economy is largely dependent upon trade. Much of Singapore's success can be linked to government policy, which has been driven to make their port a major part of the economy. Since Singapore experiences one of the highest standards of living in the world, many residents are able to live with this government control and are willing to sacrifice some civil liberties.

Australia: New Links to Asia

Australia is rich in natural resources and remains a major world exporter of rural commodities and minerals. Although the interior is resource rich, its environment is harsh and most people choose to live on the coast. Australia's population distribution is largely concentrated in a few large cities on the coastline. Australia is one of the most urbanized places in the world and one in five Australians live in the biggest city of Sydney. Rural wealth created jobs in urban areas and helped develop cities into business centers. Historically, Australia has had the closest ties with Europe and the United States. In recent decades however, Australia has been taking advantage of their spatial location and tapping new markets in Asia. This shift is also seen in the cultural landscape as the number of immigrants from Asia has increased. Australia is a perfect example of how geography shapes nations.

DISCUSSION OF UNIT THEMES

Southeast Asia's rich cultural heritage left a wealth of historic buildings and traditional ways of life that formed the basis of growing tourist industries in the 1990s. Indonesia is the most populous country, but it is geographically divided into over 13,000 islands and 300 ethnic groups based on language. Places such as Bali must find ways to balance their unique cultural traditions with the influx of cultural attitudes brought by tourists (*Contemporary World Regional Geography*, Bradshaw/White/Dymond, © 2004, McGraw-Hill Higher Education).

Today, Singapore is a multicultural and multinational global city-state with international business services to rival Hong King and major cities in Europe and the United States. The global connections, however, have to take into account local pressures. Singapore's existence and prosperity is still resented by Malaysia. Malaysia became noticeably more prosperous during and since the mid-1980s. More people moved into the towns, especially the capital, Kuala Lumpur, because of the jobs in government offices, businesses, the greater range of shopping, and increased leisure activities. Malaysia is a multiethnic society, with many people of mixed parentage. Malays, who are mainly Muslims, retain control of parliament and often regard themselves as first-class citizens and others as second-class citizens. The Chinese came to work in the mines during British rule and are now mainly middle class with commercial interests. Ethnic and cultural relations remain a high priority as Malaysia's economy continues to grow (*Contemporary World Regional Geography*, Bradshaw/White/Dymond, © 2004, McGraw-Hill Higher Education).

Australia is the only country that is also a continent. Until the mid-1990s, most Australian immigrants came from the British Isles. To increase its population and maintain links with the United Kingdom, Australia encouraged immigration by paying their costs to immigrate until after World War II. Australia's immigration policies have changed in recent decades and more immigrants are coming from Asia as Australia builds new trade relationships (*Contemporary World Regional Geography*, Bradshaw/White/Dymond, © 2004, McGraw-Hill Higher Education).

TEST YOUR KNOWLEDGE

Multiple Choice and True/False Questions

1. Indonesia has approximately how many islands?

   a. 13,000
   b. 2,000
   c. 500
   d. 10

2. Australia is building new relationships with which part of the world?

   a. South America
   b. Asia
   c. Canada
   d. Middle East

3. Which of the following best describes Australia's population pattern?

   a. evenly distributed across the country
   b. concentration in the interior
   c. concentration in a few coastal cities
   d. concentration in the northern half of the country

4. The capital of Malaysia is:

   a. Kuala Lumpur.
   b. Bali.
   c. Jakarta.
   d. Singapore.

5. Singapore's economy is largely based on:

   a. trade.
   b. tourism.
   c. natural resources.
   d. a & b

6. Unlike the rest of Indonesia, Bali's population is predominately:

   a. Hindu.
   b. Buddhist.
   c. Muslim.
   d. Christian.

7. Vietnam's most sustainable export is:

   a. coffee.
   b. cocoa.
   c. rice.
   d. tea.

8. Laos' most lucrative natural resource is:

   a. oil.
   b. water.
   c. timber.
   d. gold.

9. True/False Malaysia is cultural and ethnically heterogeneous.

10. True/False Laos is Southeast Asia's only landlocked country.

Short Answer and Essay Questions

1. How is Australia taking advantage of their spatial location?
2. Discuss the challenges that Laos faces as a communist landlocked country.
3. What factors led Vietnam to becoming the world's second largest rice exporter?
4. How are the Balinese maintaining cultural integrity in the face of tourism?
5. How has Malaysia maintained peaceful ethnic relations?

Map Exercises

1. Place the following physical features on the outline map: Australian Outback, Mekong River.
2. Using the blank outline map, make a dot for the location of the following locations: Bali, Vietnam, Laos, Malaysia, Singapore.

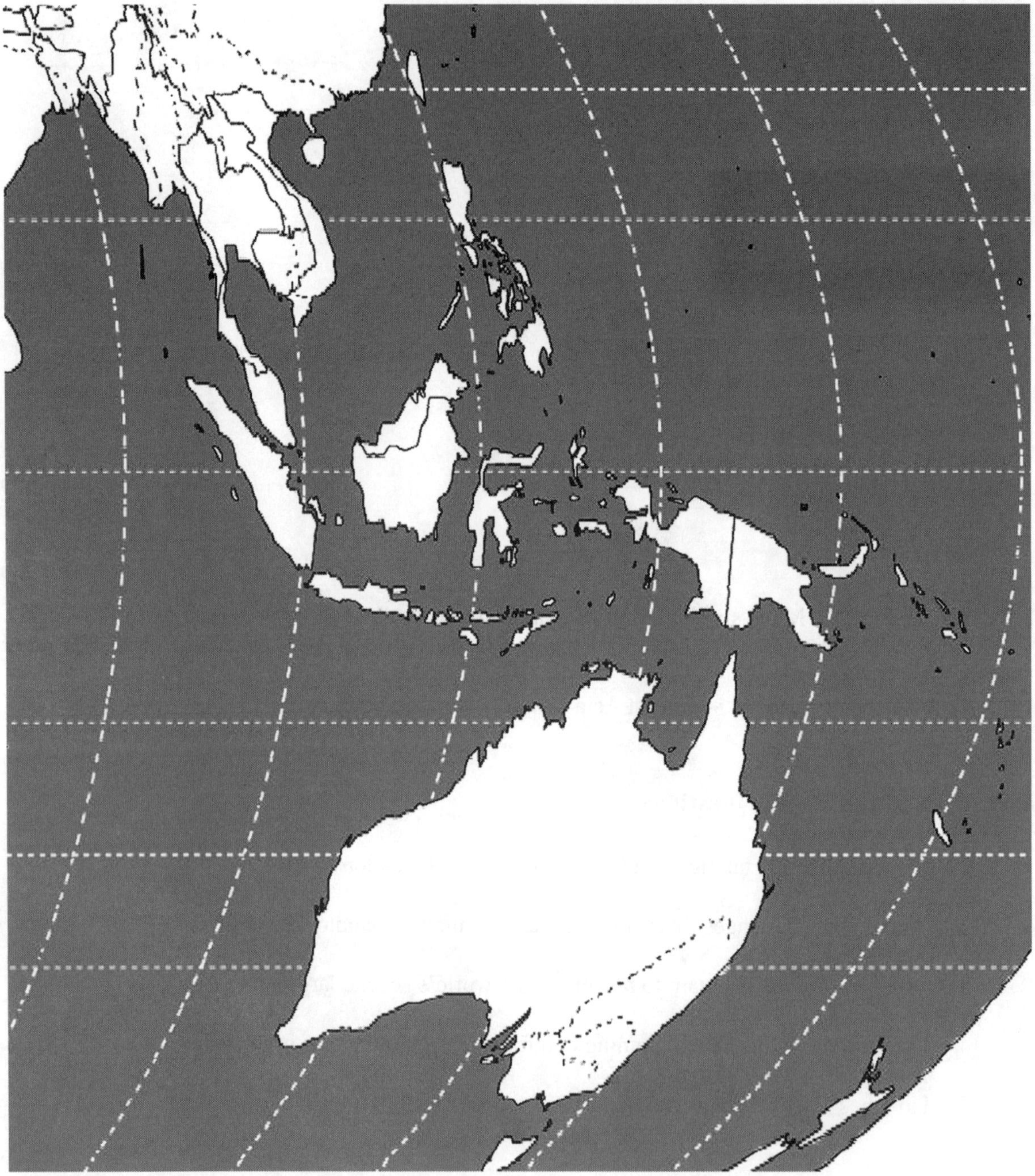

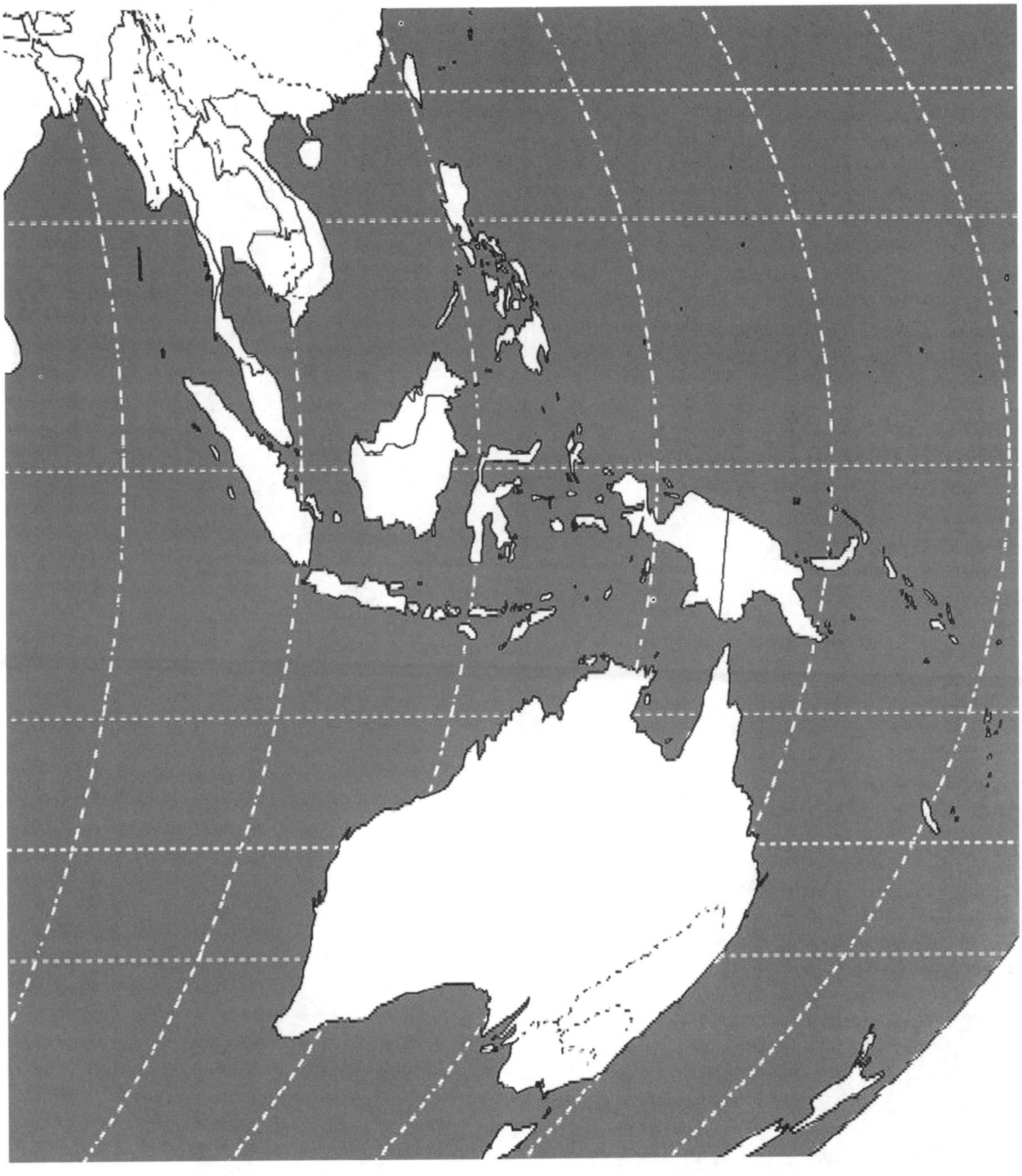

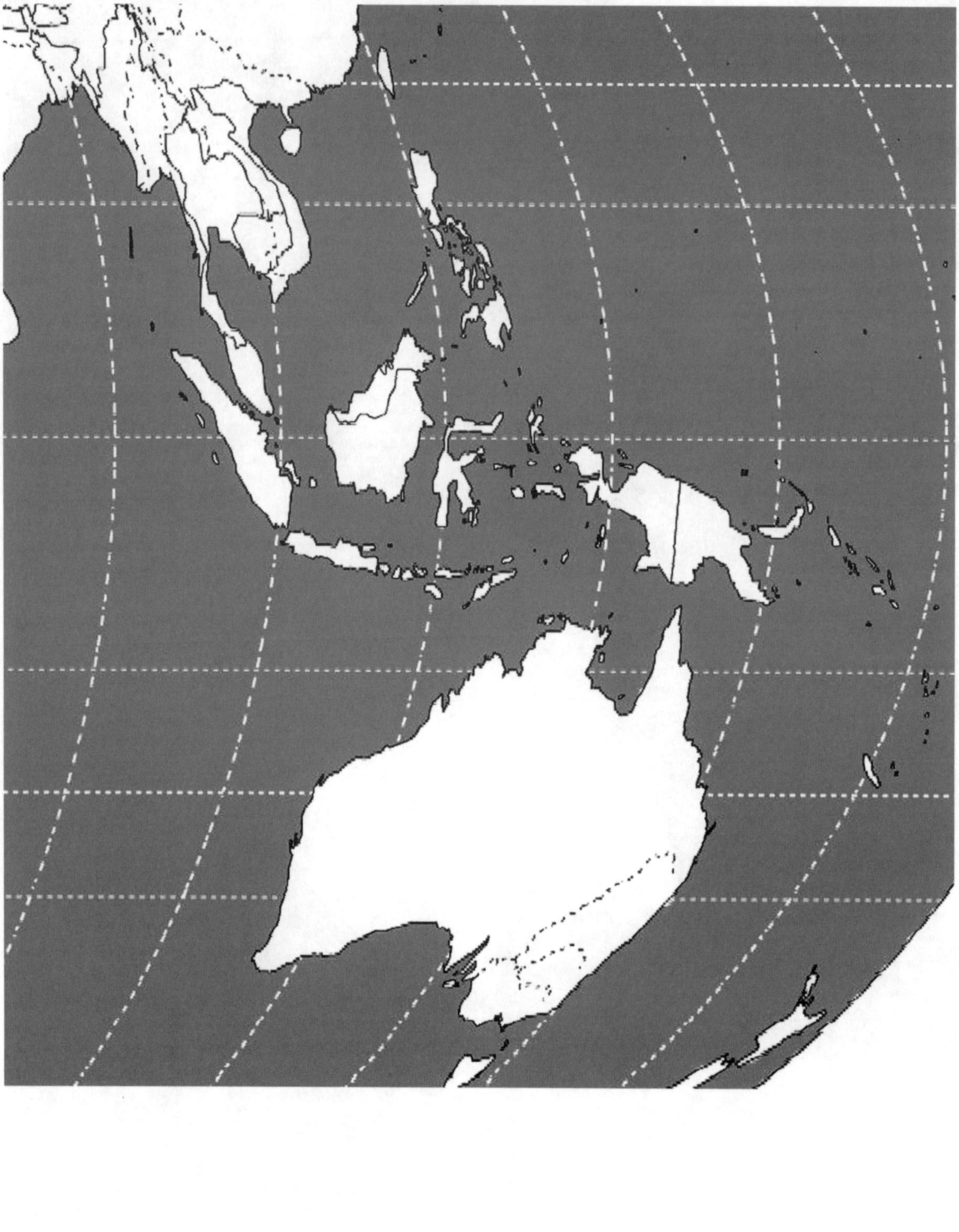

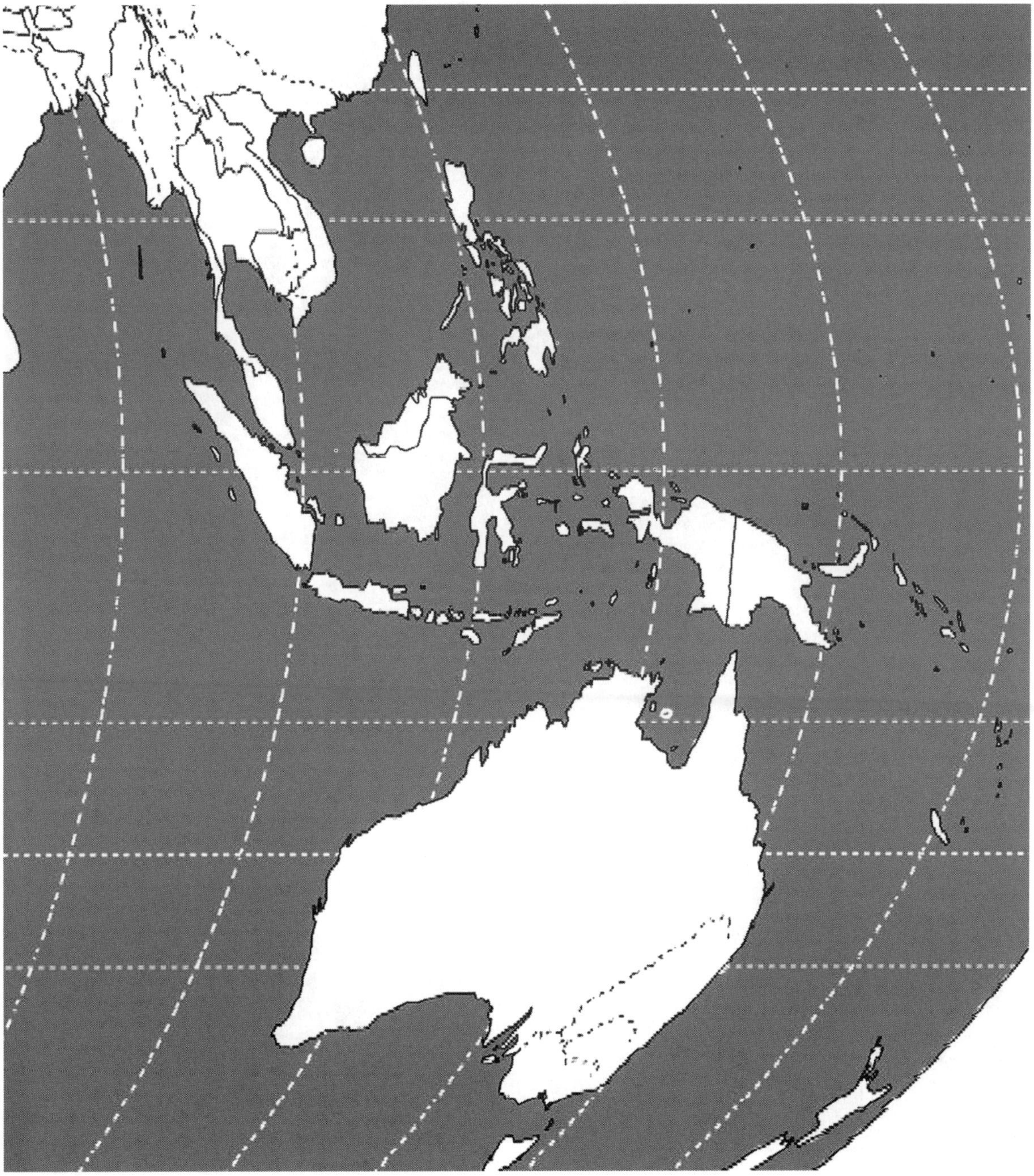

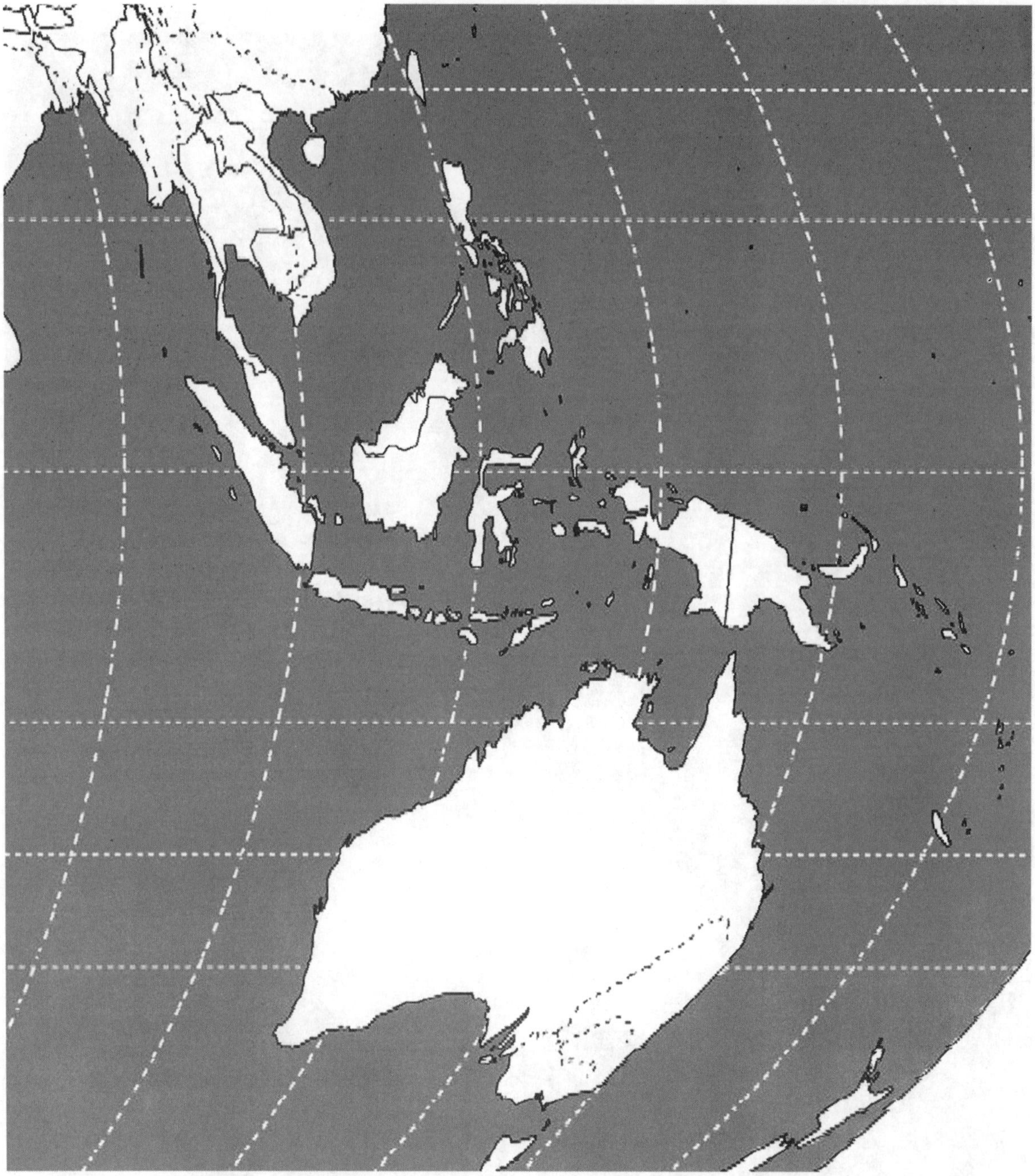

UNIT SIX: SOUTH ASIA

Unit Objectives

At the completion of the unit, you should be able to:

- understand the challenges of sustainable development.
- understand push and pull factors of migration.

PROGRAM 16

URBAN AND RURAL CONTRASTS

Key Themes

- **Rural to urban migration**
- **Metropolitan growth**
- **Water management**

Delhi: Bursting at the Seams

Delhi, the capital of India, has long attracted many people from other parts of India in search of jobs and a better life. Its current population is now booming at approximately 14 million. Old Delhi is reminiscent of an Islamic era, whereas New Delhi reflects British colonial rule. Many who come to Delhi in search of work are disappointed to discover that the unemployment rate is already very high. The problem however, is that work is dwindling in rural areas due to mechanized agriculture so many are left without work in either place. Many are left waiting for work while living in squatter towns. Squatter towns have grown to house more than one third of Delhi's population. Although city services such as sewer and garbage are not available to squatter settlements, many long time residents have managed through illegal connections. Air pollution in Delhi is at all time highs as more people bring more cars and factories. The National Capital Region plan may help. This plan will extend Delhi's city limits to 20 times its current size. Satellite cities within the region will provide the same employment, education, and health opportunities provided by the city center. While still in its planning stages, it has already proved to be somewhat successful.

Dikhatpura: Help through Irrigation

In the rural countryside of India, many people rely on traditional methods of agriculture that are often inefficient. While modern agricultural methods are slowly being introduced, sustainable development has been challenging. One third of India is subject to drought and farmers must rely on rain brought during the monsoon season. Irrigation projects began in the 1940s to help combat this issue. It is questionable however, whether or not these new agricultural techniques have led to sustainable development in the region. With irrigation water, many farmers can now grow two crops instead of one and many switched to growing cash crops instead of subsistence crops during the government's promotion of the "green revolution." This causes problems for the small scale farmer who cannot afford what is needed for cash crops such as fertilizers, pesticides, herbicides, and high yielding seeds. In addition to this "green revolution," the government has also promoted water buffalo husbandry as a way to combat poverty. This successful program is sometimes called the "White Revolution."
Water management is the key to successful irrigation practices in the area. Unfortunately water rights are not properly managed and lead to frequent disputes over use of the water, leaks in the system, and increased salinization. Critical to sustainable economic development are the issues of population growth and gender equality. Female education may prove most successful in dealing with these issues.

DISCUSSION OF UNIT THEMES

Over-population is an obvious concern for India. India's population is expected to exceed China's in the very near future. Although predominantly a rural population, more and more people are flocking to the cities in search of work and many are forced to live in squatter settlements and slums. These slums vary in type and organization. Many have no title to housing or rights to establish businesses. In an attempt to help with this problem, Delhi is planning to significantly expand its current city limits (*Contemporary World Regional Geography*, Bradshaw/White/Dymond, © 2004, McGraw-Hill Higher Education).

In the rural areas, many farmers are losing their livelihood in the wake of mechanized agriculture, further increasing rural to urban migration. The Green Revolution began in 1966 when high-yielding varieties of wheat and rice were introduced as part of Indian agricultural policy. While it did boost the agricultural economy, benefits from the Green Revolution were geographically and socially uneven as was seen in Dikhatpura. Water management may help with this problem (*Contemporary World Regional Geography*, Bradshaw/White/Dymond, © 2004, McGraw-Hill Higher Education).

TEST YOUR KNOWLEDGE

Multiple Choice and True/False Questions

1. Which of the following is not a reason why many rural people move to Delhi?

    a. education
    b. better health facilities
    c. better air quality
    d. a perception of better job opportunities

2. What has contributed to a loss of agricultural jobs in rural areas of India?

    a. conversion to mechanized agriculture
    b. lack of labor
    c. persistent droughts
    d. excessive migration to rural areas

3. The "White Revolution" refers to the husbandry of which animal?

    a. water buffalo
    b. sheep
    c. goats
    d. horses

4. True/False — The Green Revolution has proved advantageous to large plantation farmers, as well as small lot farmers.

5. True/False — India's population is expected to stabilize and then decrease within the next 25 years.

Short Answer and Essay Questions

1. Describe the impacts of the Green Revolution on small farmers in India.
2. How can gender equality help combat poverty in India?
3. What are some of the environmental consequences of Delhi's population growth?
4. Why are rural Indians moving to Delhi?

Map Exercises

1. Using the blank outline map, locate the following: India, Delhi, the Ganges River.

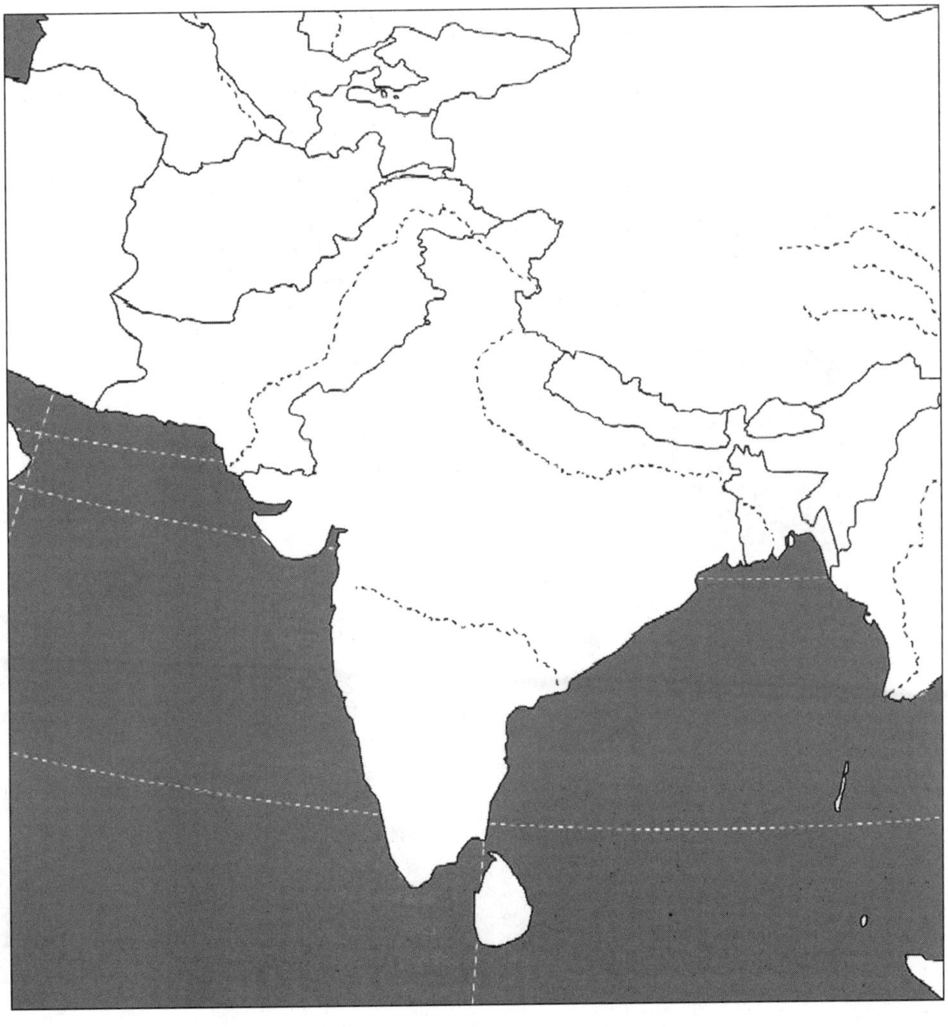

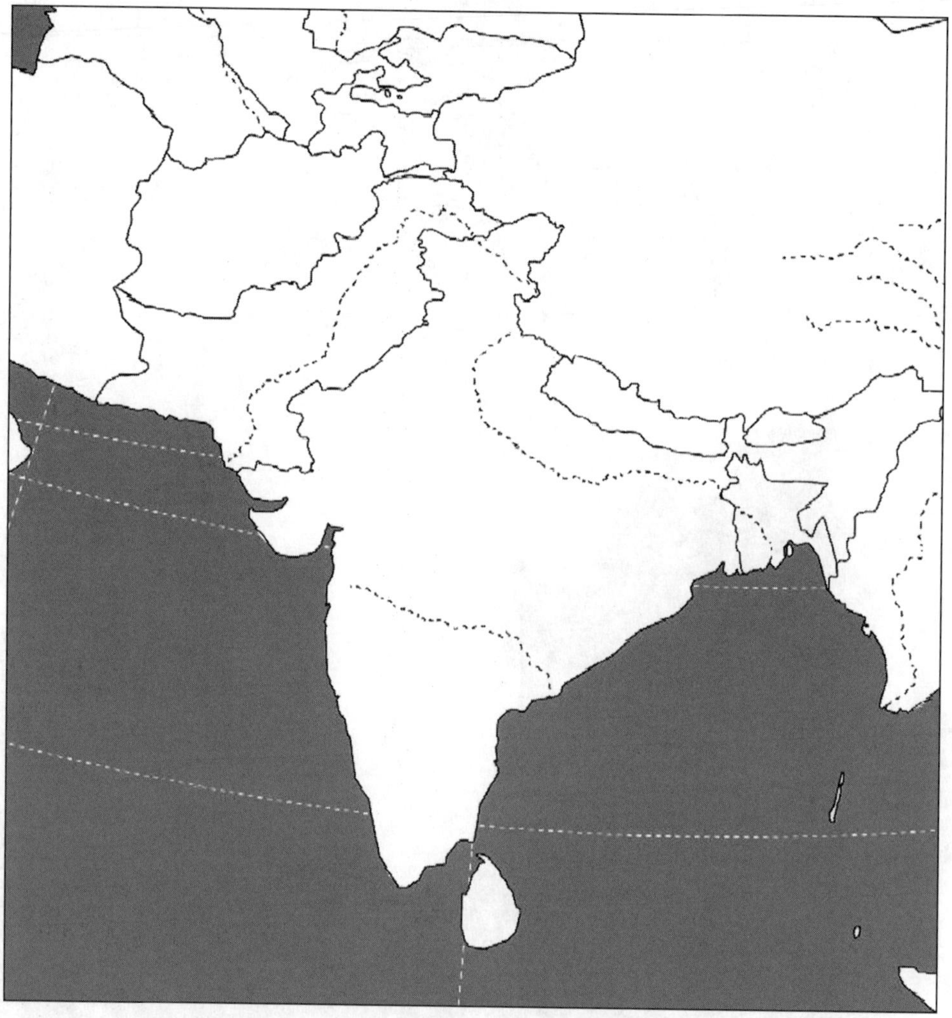

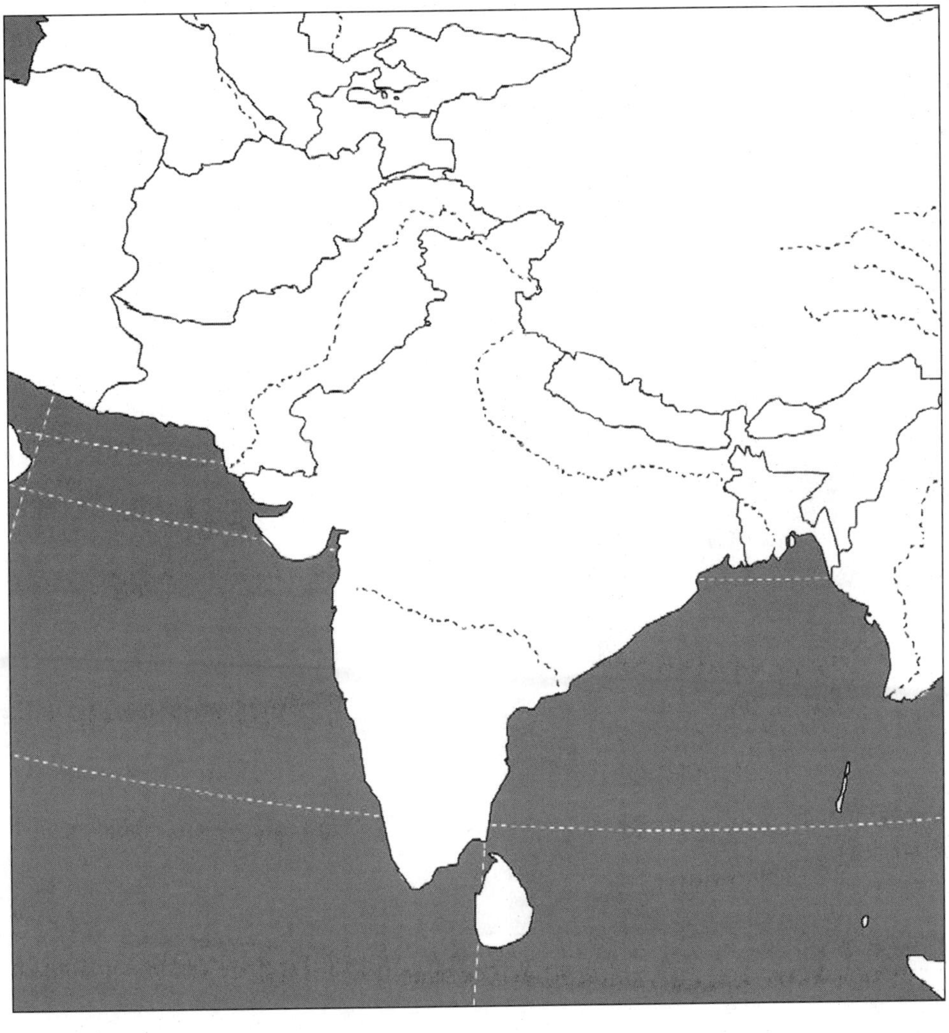

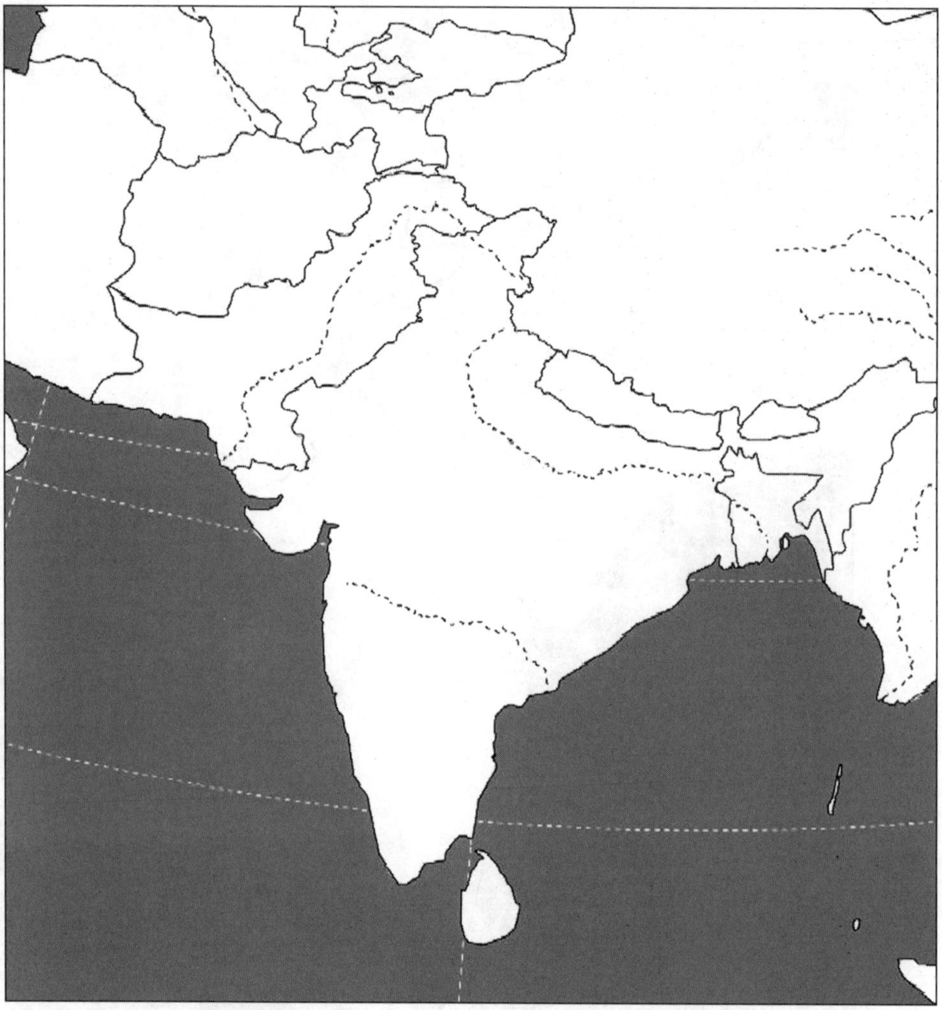

UNIT SEVEN: NORTHERN AFRICA AND SOUTHWESTERN ASIA

Unit Objectives

At the completion of the unit, you should be able to:

- understand the unique political geography of Israel.
- understand the influence of fundamentalism in politics.
- understand the differences between a secular and non-secular state.
- understand the importance of resource management.
- understand the impact of environment and population on resources.

PROGRAM 17

SACRED SPACE, SECULAR STATES?

Key Themes

- **Territorial conflict**
- **Urban landscape**
- **Fundamentalism**
- **Political geography**

Jerusalem: Capital of Two States?

The conflict in the Middle East is driven by its political and historical geography. Jerusalem lies at the heart of this conflict. Muslims, Christians, and Jews all lay claim to this city. Plagued by centuries of colonial rule, the region does not have a long history with concrete political borders. It was largely a religious city having extreme significance for both Muslims and Jews. Conflict erupted after the British withdrew. Territory was now up for grabs. The United Nations proposed dividing Palestine into Palestinian and Jewish states, but seen as unfair for a variety of reasons, the Palestinians rejected it. Israel proclaimed statehood regardless, the fighting began, and the Jews began enlarging their territory, including the western half of Jerusalem. After the victory of the six-day war in 1967, Israel moved its capital to Jerusalem from Tel Aviv and redrew the city boundaries. Conflict continued between Israel and the Palestine Liberation Army. In 1993 an agreement of mutual statehood seemed promising, however the issue of capitals remained at the forefront. Palestine wanted Jerusalem as their capital. A divided city seemed like the only solution. Negotiations began to split the city along ethnic lines, but the oldest part of the city was still debated. The relatively small "Old City" contains some of most historically significant and religious buildings for both ethnic groups. Various negotiations were rejected by both parties. It is believed by many that in order for peace to exist in the region there will need to be two states and a divided Jerusalem.

Turkey: Fundamental Change

Turkey's location at the edge of the European realm poses advantages and disadvantages. Close proximity to the rest of Europe gives Turkey an economic advantage, however increasing Islamic influence may culturally strain the country. Istanbul is the largest city in Turkey. Most residents are Muslim, however there are wide contrasts in lifestyles. Many see Islam playing a limited role in government, while others feel that Turkey should be an Islamic state. Still others see Islam playing a key role in local government rather than national. The conflict has led to the rise of fundamentalist political groups. Istanbul is at the heart of this issue because of its advantageous location at the crossroads of Europe and Asia. Economic modernization has led to an increase in rural to urban migration and Istanbul's population has grown tremendously in the past few decades. To many, increased economic might requires maintaining a secular state. Political obstacles however, are keeping Turkey from fully integrating into the European Union.

PROGRAM 18

OIL AND WATER

Key Themes

- **Population growth rate**
- **Exponential population growth**
- **Omanization**
- **Resource development**

Egypt: Gift of the Nile

The Nile runs 4,000 miles through some of the driest terrain of Africa. Its bank and delta contain some of the most fertile areas in the world. The city of Cairo depends on this river for its survival. The ever-expanding population is outgrowing the agricultural ability of the region, and there is pressure to use more of the Nile's water for desert irrigation. The Aswan High Dam and Lake Nassar gave Egypt the ability to tame the Nile's water for agriculture on the delta. Despite the advantages of the dam, there are concerns with silting and salinization which depletes soil nutrients. Since the 1990s, land reform policies have shifted the once socialist system of agriculture to a capitalist system, putting small farmers at a disadvantage. Economically however, it has been quite successful and the program has expanded into new areas of the desert. It is not only agriculture that is claiming desert land, but urbanization and urban growth is taking over some of the best farmland. Egypt was forced to find new ways of using the desert. In 1997 they began a series of projects, which will ultimately create a new delta parallel to the existing one. Questions exist as to whether there will be enough water, especially in the face of resource use conflicts with neighboring countries.

Oman: Looking Beyond Oil

Oman, along with many other countries in the Middle East, is rich in oil. Some fear that this resource will not last forever and Oman has taken a unique initiative to reduce its dependency—educating and employing women. Modern societies exist in Oman because of its lucrative resource and now Oman aims to diversify its economy. Oil production costs tend to quite high and Oman's output is low compared to its oil-producing neighbors. It is expected that their reserves will run out first. Diversifying the economy requires educating people so they can move into private industry sectors. Educating women is one way to help this transition since the security of the public sector makes the private sector less attractive for most natives. It is now common in Oman for women to work and have a family as well. The status of women has also grown and many hold positions of authority. Oman is also striving for "Omanization" of the workforce, in other words, training Oman natives to do work that has traditionally been done by guestworkers. Omanization and diversity can very well hold the key for a bright future if and when the oil runs dry.

DISCUSSION OF UNIT THEMES

Israel is a unique example of a country created by the United Nations for a particular religious group, despite opposition from those living in and around it. As seen in the Power of Place, Jerusalem presents a major problem for Israeli-Palestinian negotiations. After Israel took East Jerusalem in the 1967 war, it confiscated land in the occupied area and made Palestinians second-class citizens by denying them property rights. Israel proclaimed Jerusalem to be its capital in 1950 although the United States and most other countries do not recognize Jerusalem as the Israel capital and maintain embassies in Tel Aviv (*Contemporary World Regional Geography*, Bradshaw/White/Dymond, © 2004, McGraw-Hill Higher Education). Many believe that there will not be peace in the region until there are two separate states and a divided Jerusalem.

Political geography and fundamentalism also play a role in Turkey. After the defeat of the Ottoman Empire in World War I, Turkey became a nationalist and secular republic, putting the country before religion in questions of government. Turkey gradually modernized and became increasingly involved in the global economy. Turkey faces continuing challenges from Islamic political groups to its long-term secular principles. When a militant Islamic group was elected to government in the late 1990s, the military took over and installed a secular group into power (*Contemporary World Regional Geography*, Bradshaw/White/Dymond, © 2004, McGraw-Hill Higher Education).

Water resources are extremely important in the dry regions of North Africa and Southwest Asia. The major civilization of the Nile and Tigris-Euphrates River valleys were based on irrigation water, and there were many local developments bringing water from its source to areas of use. The competition for use of the Nile River waters demonstrates that political and social factors are as important as environmental in the economic development of natural resources. The arid environment deepens the problem. Irrigation farming in arid areas requires good management to slow down the rate of salinization (*Contemporary World Regional Geography*, Bradshaw/White/Dymond, © 2004, McGraw-Hill Higher Education). Egypt's ever expanding population is contributing to the problem. This exponential population growth rate is far exceeding the arithmetic growth rate of its resources and is forcing Egypt to find new ways of utilizing the Nile's waters, such as building an entirely new delta.

Oman also is taking resource management under careful consideration—the resource being oil. Oman is located on the Persian Gulf and has the advantage of a long Indian Ocean coast, linking it to eastern Africa. Although the 2001 population of Oman was only 2.4 million people, it is globally connected and is comparably progressive in the Arab world (*Contemporary World Regional Geography*, Bradshaw/White/Dymond, © 2004, McGraw-Hill Higher Education). Omanization is an attempt to increase employment of Omanis in jobs that have been traditionally held by guestworkers. Employment and education of women is also encouraged. Together these initiatives may help Oman in the future should its oil resources run low.

TEST YOUR KNOWLEDGE

Multiple Choice and True/False Questions

1. Which three religious groups occupy Jerusalem?

   a. Jews, Christians, Hindus
   b. Moslems, Jews, Buddhists
   c. Jews, Christians, Moslems
   d. Hindus, Jews, Buddhists

2. Which area of Jerusalem is most debated?

   a. West bank
   b. Old City
   c. East bank
   d. Golan Heights

3. Which dam is located on the Nile River?

   a. Aswan High Dam
   b. Hoover Dam
   c. Three Gorges Dam
   d. none of the above

4. Which of the following is the world's largest man made lake?

   a. Lake Nasser
   b. Lake Victoria
   c. Lake Michigan
   d. Lake Superior

5. Currently, Oman's largest export is:

   a. natural gas.
   b. coffee.
   c. oil.
   d. cocoa.

6. Which of the following best describes Omanization?

   a. increased employment of Omanis in what are now heavily guestworker sectors
   b. increased employment of guestworkers to help with the lack of labor
   c. hiring only women in public sector jobs
   d. increasing growth of the public sector

7. True/False Oman's oil production is much lower than that of neighboring Middle Eastern countries.

8. True/False Egypt is currently working on building an entirely new delta parallel with the current Nile delta.

9. True/False Turkey is a secular state.

10. True/False The city of Jerusalem is predominantly Arab.

Short Answer and Essay Questions

1. Describe the debate over Jerusalem's Old City?
2. What challenges does Turkey face in joining the EU?
3. What is unique about the spatial location of Turkey?
4. Why is Egypt so dependent on the Nile River?
5. What role do women play in the economic development of Oman?

Map Exercises

1. Using the blank outline map, find the following locations: the Nile, Oman, Israel, Jerusalem, Turkey, Istanbul.

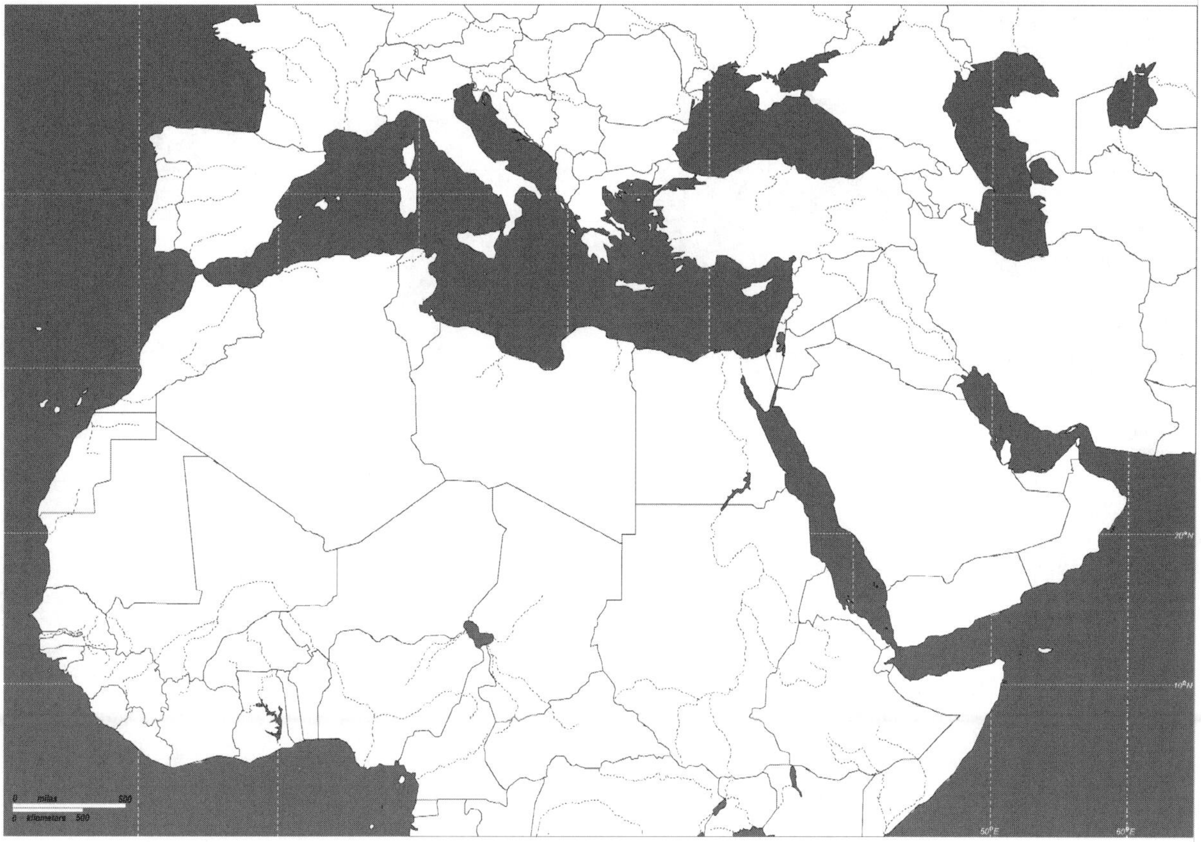
0 miles 500
0 kilometers 500
20°N
10°N
50°E
60°E

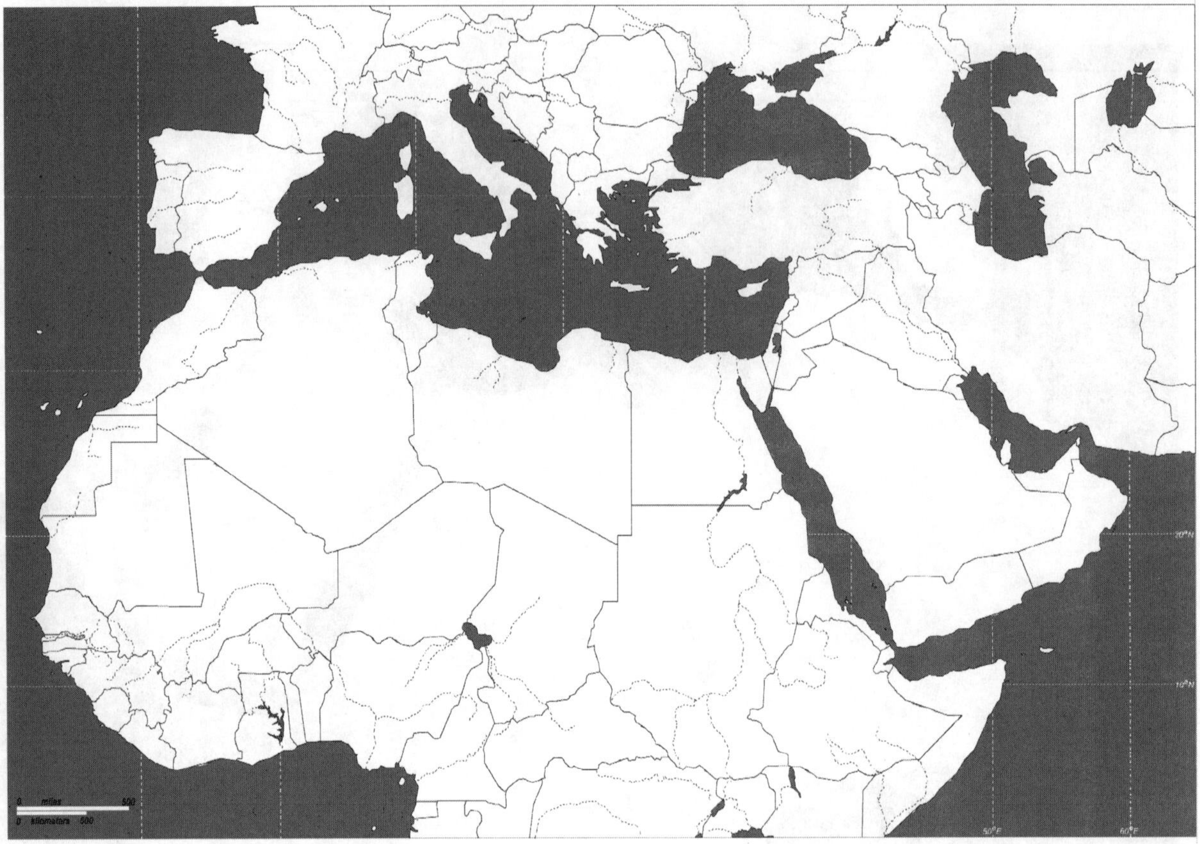

20°N
10°N
0 miles 500
0 kilometers 500
50°E
60°E

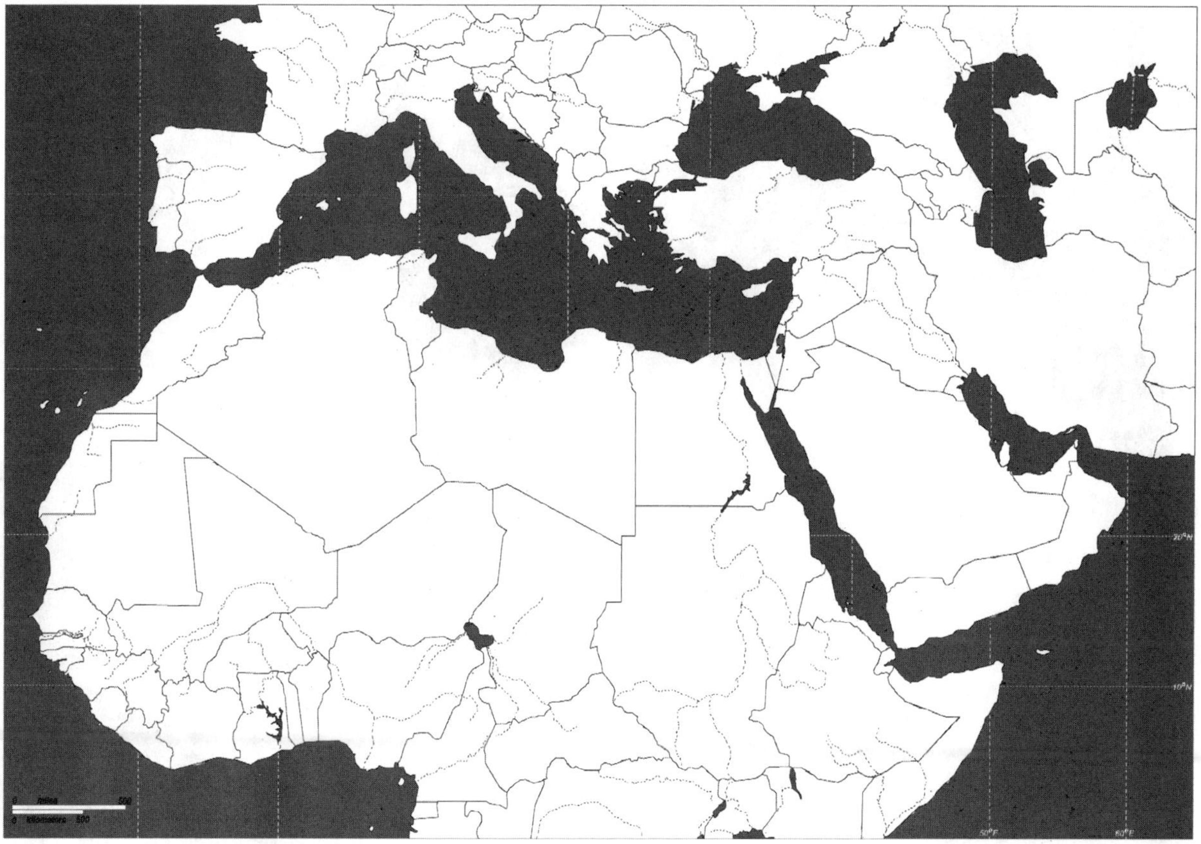

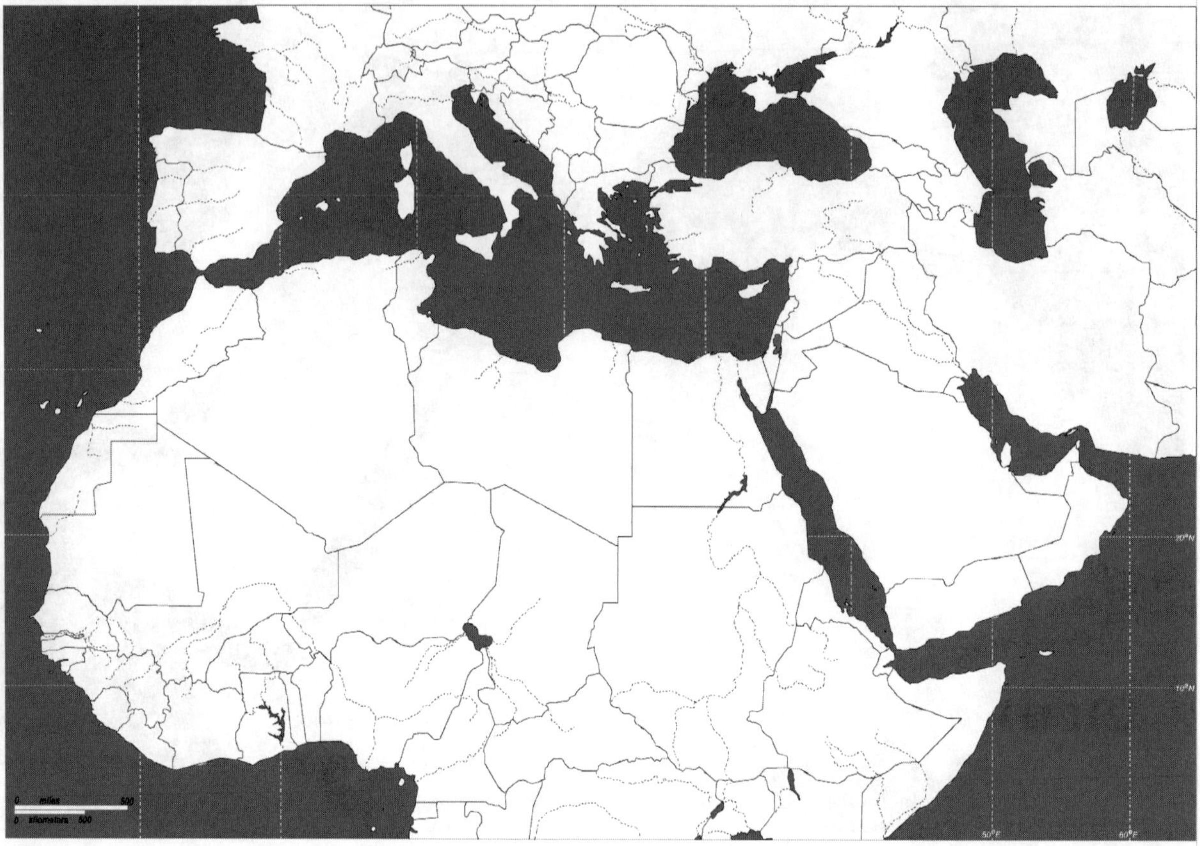
20°N
10°N
0 miles 500
0 kilometers 500
50°E
60°E

UNIT EIGHT: AFRICA SOUTH OF SAHARA

Unit Objectives

At the completion of the unit, you should be able to:

- understand the process of disease diffusion.
- understand the political processes behind land inequities.
- understand relationship between colonialism and resource use.

PROGRAM 19

STRENGTH TO OVERCOME

Key Themes

- **Land reform**
- **Medical geography**
- **Disease diffusion**

South Africa: This Land is my Land

During apartheid, South Africa's black population was moved onto homelands—marginal, arid areas with little infrastructure. The white minority controlled much of the arable land and grew profitable cash crops, while blacks barely grew enough for subsistence agriculture. After the end of apartheid, one of the biggest challenges was land reform. New land reform policies attempted to undo years of inequality. While land is being returned to blacks, production is slow and training is needed to ensure the success of farmers.

Kenya: Medical Geography

Life expectancy in Africa is the lowest on the earth. While many variables such as poverty, famine, and war contribute to this problem, disease has had a most devastating effect. Many diseases plague Africa. Malaria, sleeping sickness, cholera, and ebola are just a few of the diseases that are common. One of the biggest killers of recent years is AIDS and Kenya has been hit particularly hard. It is estimated that over two million people in Kenya are infected. Upon mapping the diffusion of the disease in Kenya, it was discovered that AIDS tends to diffuse hierarchically, hitting the largest urban centers first, as well as contagiously from the source region. The spread of the disease has been attributed to a weak infrastructure, rural to urban migration, and sexual practices. Aggressive campaigns to increase awareness have proved quite successful in many African countries including Uganda and Senegal. If the campaigns continue to succeed, Africans may also be able to combat other deadly diseases that have hampered Africa's economic potential for decades.

PROGRAM 20

DEVELOPING COUNTRIES

Key Themes

- **Subsistence agriculture**
- **Primary resources**
- **Sustainable development**

Cote d'Ivoire: Cocoa and Change

Cote d'Ivoire is the largest cocoa producer in the world. Plantations are located all along the southern area of the country and people migrate from regions near and far to work on them. It is estimated that as much as 40% of the population are foreign. The stabilization fund once controlled the agricultural economy in Cote d'Ivoire. Each year the state would fix the price and keep any excess should the world price exceed it. This helped many at the top of the economic scale to get richer and widened the gap between the rich and poor. The dichotomy is evident is the urban landscape. The infrastructure improved with the help of export money and allowed economic expansion into the northern areas of the country. The capital was moved to the more central location of Yamoussoukro to help rebalance the country. In 1986, the Cote d'Ivoire economy suffered when world cocoa prices plummeted. After 1994 however, world market prices improved and the government decided to eliminate the stabilization fund and have merchants deal directly with the farmers. On the positive side, production has increased, however increased production and cultivation is damaging the forests. Natives are finding less and less land to live off of and tensions are mounting between native Ivorians and multiple generation foreigners and for the first time, Cote d'Ivoire began experiencing political instability. Political instability led many aid donors to stop dispersing aid and Cote d'Ivoire fell into another recession in 1999. The recession continues although some donors have come back. The future is uncertain, but many hope for positive growth rates.

Gabon: Sustainable Resources?

Most of Gabon's income is generated by the exportation of raw materials and most of the activity in central Gabon involves exploitation of the forest. Although oil revenues have slowed down the timber industry, the forest still provides 15% of the country's jobs. The Transgabonese Railway was completed in 1986 and it was thought it would serve to help develop the interior part of the country, however it primarily serves to pull resources (particularly timber) from the interior to the coast. In 1957 oil was discovered in the Port-Gentil Zone and oil now makes up 80% of the country's exports. Increased oil revenue led to an increase in imports and the country imports 50% of its food requirements. Gabon slowly worked to reduce its share of the oil sector and increase shares in other sectors such as agriculture and forestry. This shift however, has increased exploitation of the forest and it is predicted that Gabon will lose half its forest cover in the next 100 years. In 2002, the Gabonese president established a new natural park system that covers 10% of Gabon's rainforest. This has also paved the way for ecotourism and the possibly of making Gabon's logging industry sustainable.

DISCUSSION OF UNIT THEMES

Agriculture remains the main economic occupation of over 70% of the population of sub-Saharan Africa. Most farming is for subsistence, using traditional methods to grow tropical root crops, cereals, fruits, and vegetables. Throughout the 1980s, Cote d'Ivoire diversified its established products such as cocoa, coffee, and palm oil into bananas, pineapples, cotton, and rubber. In the 1990s, the government privatized the largest rubber plantations. For a few years, growers benefited from the CFA devaluation, with exported crops earning up to twice the previous price (*Contemporary World Regional Geography*, Bradshaw/White/Dymond, © 2004, McGraw-Hill Higher Education).

Medical geography studies diffusion patterns of disease, as well as distribution patterns of health care. In 2000, nearly 4 million new HIV infections occurred in sub-Saharan Africa. Half the miners in South Africa are HIV carriers, and thousands of orphans, often carriers themselves, constitute a growing need for help in the region. Possible measures for controlling the spread of AIDS are complex and focus on treatments that recognize human frailty by providing cheap or free condoms and needles. Some countries such as Kenya and Uganda have been somewhat successful in campaigning against the disease, although rates still remain relatively high (*Contemporary World Regional Geography*, Bradshaw/White/Dymond, © 2004, McGraw-Hill Higher Education).

TEST YOUR KNOWLEDGE

Multiple Choice and True/False Questions

1. Which of the following best describes a black homeland under apartheid?

    a. predominantly black with rich soils and many service amenities
    b. predominantly white shantytowns
    c. downtown ghettos
    d. marginal areas with little rainfall and minimal infrastructure

2. The white minority controlled approximately how much of the arable land under apartheid?

    a. 85%
    b. 50%
    c. 25%
    d. 5%

3. Which of the following best describes contagious diffusion?

    a. spreading person to person outward from the source area
    b. spread from migration or relocation
    c. spreading from large cities to small cities
    d. a & b

4. Which of the following countries has seen a dramatic drop in HIV/AIDS rates due to successful government campaigns?

    a. South Africa
    b. Uganda
    c. Botswana
    d. Chad

5. Currently, Cote d'Ivoire's largest export is:

    a. bananas.
    b. coffee.
    c. oil.
    d. cocoa.

6. Gabon's timber industry grew as a result of decreasing its share of which sector?

    a. service
    b. computer technology
    c. agriculture
    d. oil

7. True/False Cote d'Ivoire is the only African country without political instability.

8. True/False South Africa no longer operates under the apartheid system.

9. True/False Malaria and sleeping sickness are still commonly found in sub-Saharan Africa.

10. True/False HIV/AIDS in Africa is only an urban problem.

Short Answer and Essay Questions

1. Discuss some the challenges South Africa faces in the process of land reformation.
2. What methods have proven successful in the fight against HIV/AIDS?
3. Why did the once politically stable country of Cote d'Ivoire experience civil unrest? What were some the consequences of this?
4. Why are diseases, which are eradicated in most of the world, still prevalent in Africa?
5. Why does Gabon rely so heavily on imports?

Map Exercises

1. Using the blank outline map, find the following locations: Kenya, Gabon, Cote d'Ivoire, South Africa.
2. Using the blank outline map, find the following cities: Libreville, Abidjan, Yamoussoukro, Nairobi, Kampala.

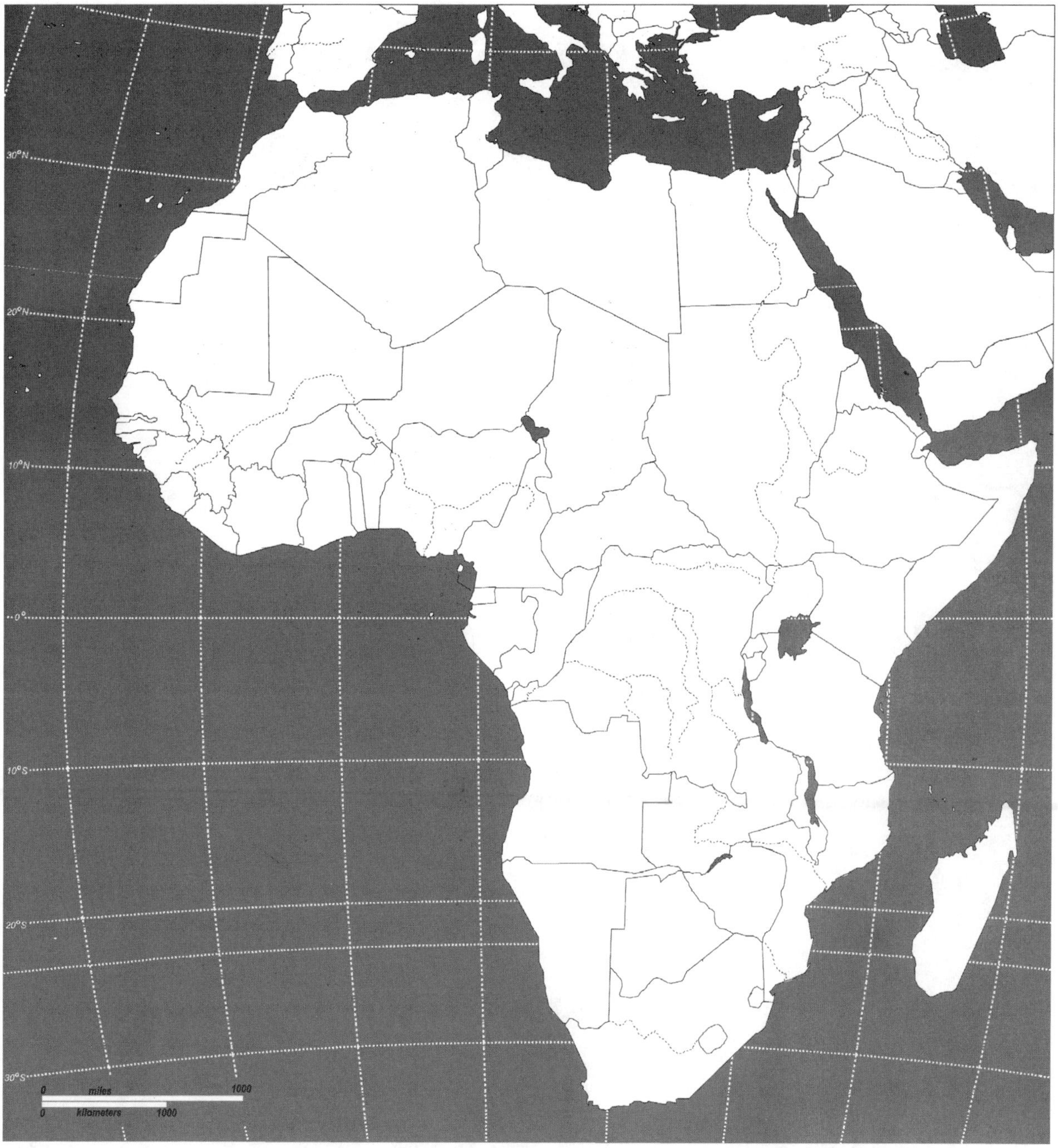
30°N
20°N
10°N
0°
10°S
20°S
30°S
0 miles 1000
0 kilometers 1000

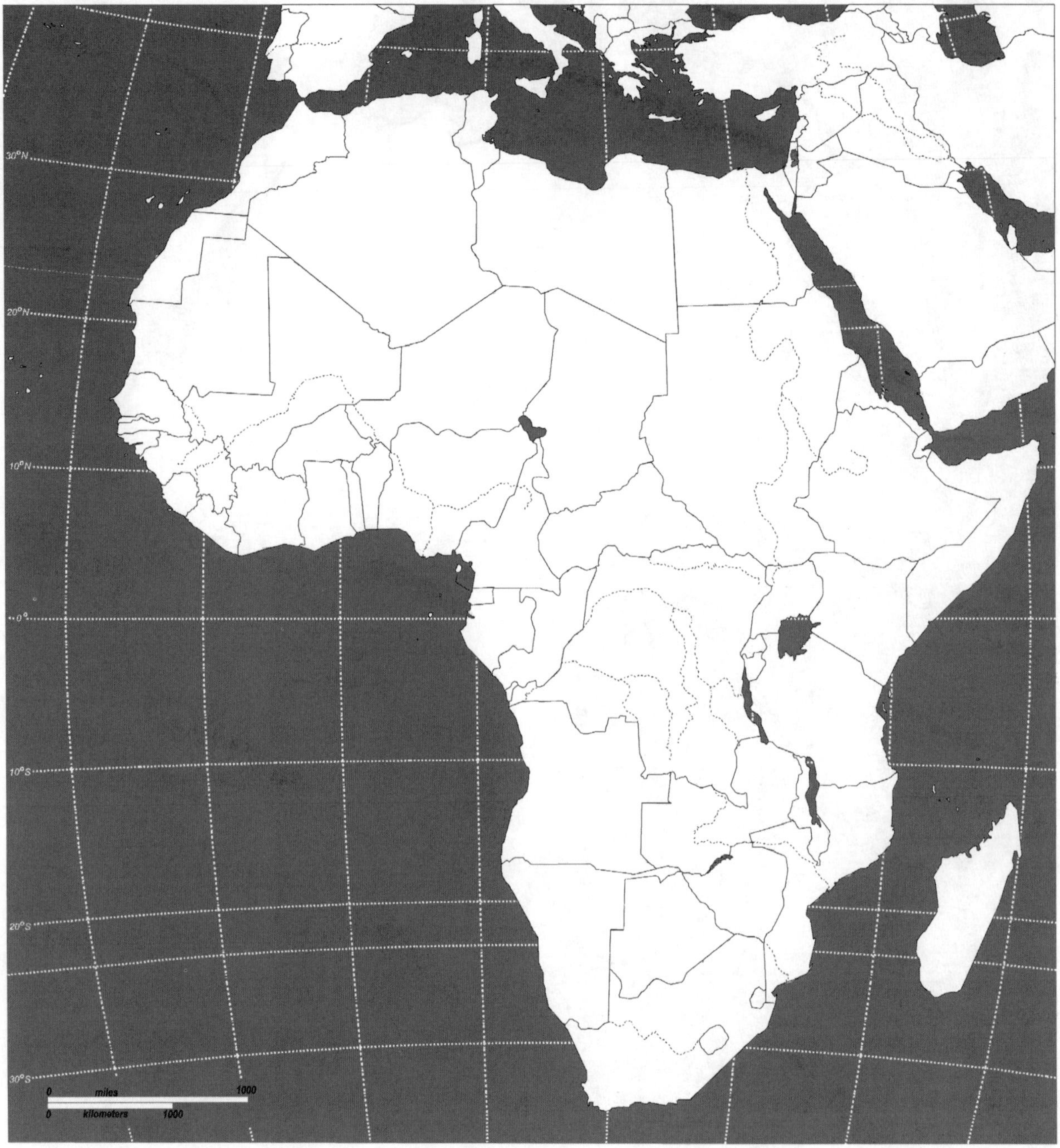
30°N
20°N
10°N
0°
10°S
20°S
30°S
0 miles 1000
0 kilometers 1000

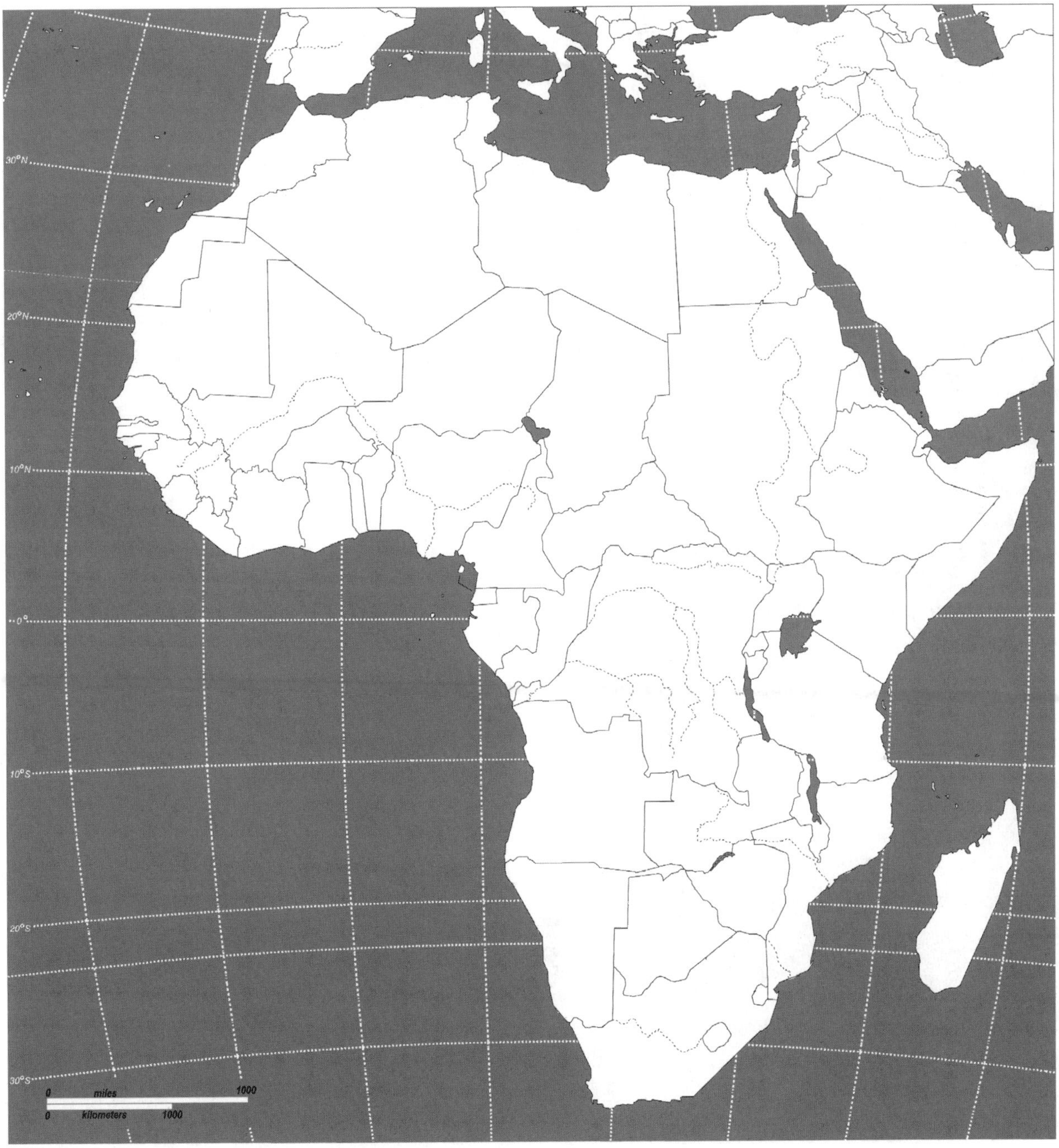

30°N
20°N
10°N
0°
10°S
20°S
30°S
0 miles 1000
0 kilometers 1000

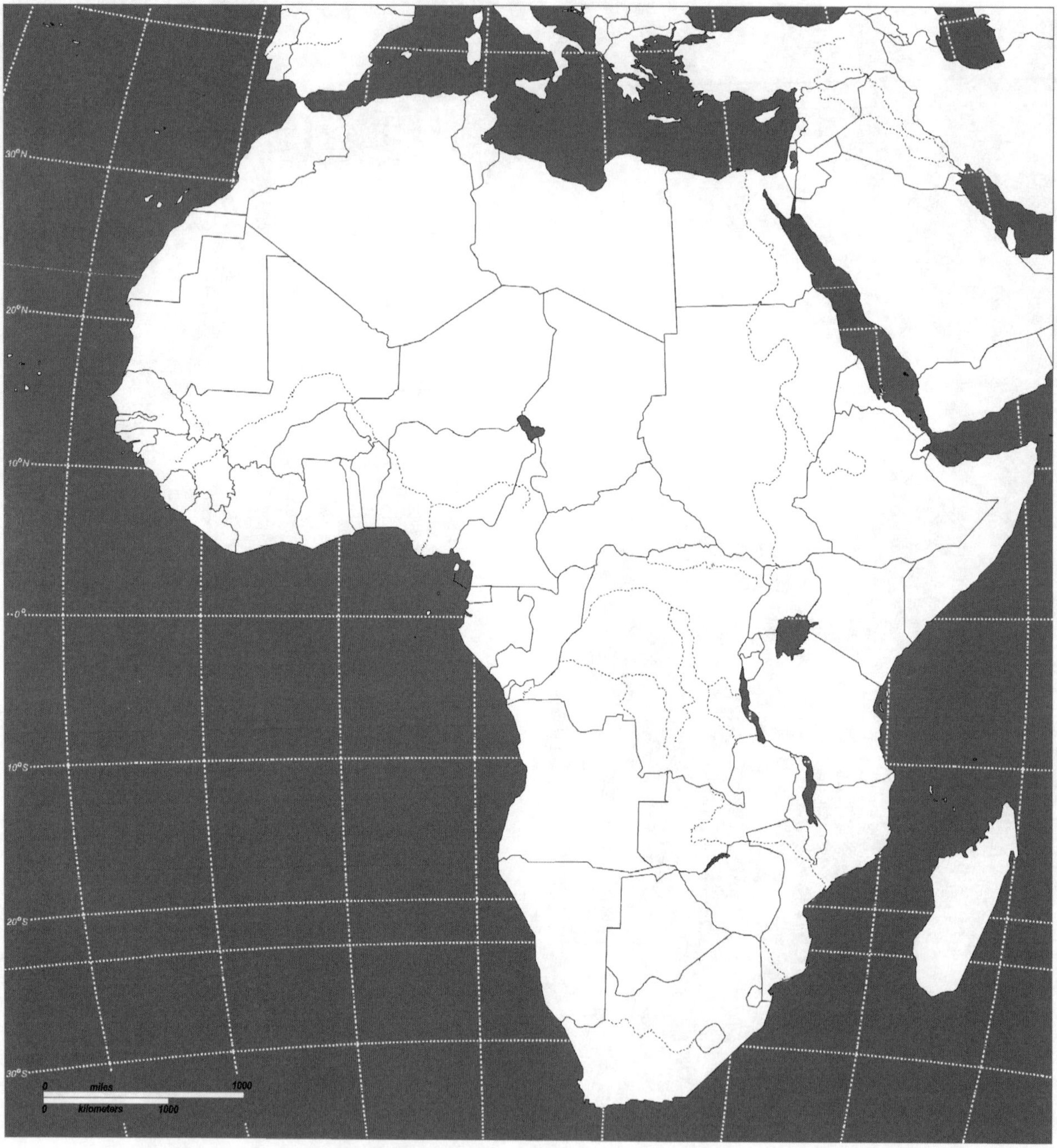
30°N
20°N
10°N
0°
10°S
20°S
30°S
0 miles 1000
0 kilometers 1000

UNIT NINE: LATIN AMERICA

Unit Objectives

At the completion of the unit, you should be able to:

- understand the forces behind internal and external migration.
- understand the process and implications of urbanization.
- describe a typical Latin American city model.
- understand spatial advantage.

PROGRAM 21

POPULATION GEOGRAPHY

Key Themes

- **Migration**
- **Push/Pull factors**
- **Population growth**
- **Resource depletion**
- **Land alienation**

Mexico: Motive to Migrate

Understanding the processes behind illegal migration from Mexico into the United States has been a challenging problem. Upon mapping the hometowns of captured illegal migrants, a clear pattern emerges that gives insight into regional economic disparities within Mexico. Most migrants come from either the northern regions or urban centers such as Mexico City. It is surprising however, that there is also a cluster of migrants from north central Mexico, an area lacking in resources and known as the "hollow core." Since many from the region have a strong sense of place, migration is only temporary, allowing enough time to save money to return home to their families. Many from the hollow core also migrate to northern Mexico to work in maquiladoras. Many maquiladoras are starting to locate farther south to take advantage of the cheaper labor from the hollow core. While at first it would seem as though this might decrease the number of illegal immigrants in the United States, tougher laws that make it riskier to cross back and forth numerous times have resulted in many Mexicans only crossing once and staying in the United States with their entire families.

Guatemala: Population and Conquest

Rural to urban migration is a common process in Latin America. In the highlands of Guatemala, an area with a long history of civil unrest, thousands of Maya have migrated to Guatemala City in search of work. During the time of the Spanish conquest in the 16th century, Indian labor was exploited. Many were forced to work in mines and plantations. The new labor force lived in "congregaciones," small densely populated labor camp settlements in the valleys. The dense living conditions, coupled with foreign diseases and unsanitary conditions decimated the Mayan population. After the collapse of the Spanish empire, the surviving Mayans returned to their original dispersed settlement patterns and their numbers have rebounded, now reaching extreme population levels. Population numbers are now beginning to outstrip the land and many are thus forced to find work in cities since agricultural land cannot support the larger families. To aggravate the situation, much of the agricultural land is held by only a few farmers, who in turn, pay the Maya meager wages for their labor. This leads to violence and civil unrest in the region. Hopefully the Mayan population explosion will begin to stabilize, perhaps with the aid of family planning. This appears to the most immediate solution for the region's problems.

PROGRAM 22

THE DYNAMIC PACIFIC RIM

Key Themes

- **Natural disasters**
- **Spatial advantage**

Ecuador: Orange Alert

The Andes region in Ecuador is home to hundreds of volcanoes, many of which are active. The town of Banos lies at the base of the Tungurahua volcano. There are a number of ways to monitor changes in the volcano that may suggest an impending eruption, however no method is guaranteed. The obvious question may be why people choose to live in the shadow of such a natural threat. For some, this may be the only land they can afford. For others, the fertile land and good weather outweigh the risk. Still others have family ties and businesses. Each person has his or her own human perception of the amount of risk involved. The challenge for researchers is to determine when to warn the public of a possible eruption. A warning could save thousands of lives, but if they are wrong it could cost the region economically. In September 1998, thousands of residents of Banos were evacuated following an orange alert. By December however, there was still no major eruption and people had been away from their homes and businesses for almost two months. After reports of looting, thousands of residents rioted and returned to their homes. No one can say when or if Tungurahua will erupt in the near future. The researchers can only predict and hope that the public will respond.

Chile: Pacific Rim Player

Chile, with its traditionally disadvantageous elongated shape, has become one of the Pacific Rim's key players. Its exports have risen dramatically in pure numbers, as well as diversification of products. Its diversity of climates and resources, coupled with its coastal location were prime ingredients to economic prosperity once Chile opened its doors to international trade in the mid 1970s. Its elongated shape provides for the ability to have multiple growing seasons and its relative location to the Andes provides a huge water reservoir for irrigation. In recent years the region has seen both employment rise and a concentration of ownership among landholders. This in turn has come to the detriment of many small local farms. Many people are flocking to the large corporate farms for employment, keeping wages low. Despite these changes, Chile continues to take advantage of its spatial location. Looking beyond the Pacific, Chile is now focusing on Asia, and Japan has become the biggest importer of Chile's wood products, keeping Chile on the global economic market.

PROGRAM 23

BRAZIL: THE SLEEPING GIANT

Key Themes

- **Urbanization**
- **Squatter settlements**
- **Latin American City Model**
- **Sustainable development**

Sao Paulo: The Outer Ring

Surrounding Sao Paulo's impressive American style skyline lie hundreds of shantytowns. Many who can't afford to live in the city center build on unclaimed land on the outskirts of the city. Sao Paulo is considered a mega city due to its large population of approximately 18 million people. It is an ethnically diverse city with many new and old immigrants claiming different regions as an expression of their identity. A wave of Italian immigrants came to Brazil in the late 1800s, Blacks moved into the city when slavery was abolished, and there was an influx of Japanese following World War II. More recently, internal immigration from Brazil's poorer northeastern states has boomed, further cramping the already congested city population. Many are forced to build their own homes and neighborhoods on the periphery of the city. This "self constructed" landscape provides much needed security for new immigrants who would otherwise not be able to afford rent. Most of these newly constructed neighborhoods are not officially recognized and therefore do not receive city services and many are built on unstable ground. The growth of these shantytowns has stretched Sao Paulo to more than 50 miles. Slowly, some neighborhoods are improving and with government negotiations are receiving minimal services. Immigration has recently slowed, but like most developing nations, a high birth rate continues to increase its population.

A Second Chance for Amazonia?

Thousands of square miles of the Amazon have been cut down for human development. For decades there has been a debate between those who want to keep the forest and those who want to develop it. In the 1970s the government started a program for Amazonia development providing incentives and road access. Cattle ranching and logging became the primary sources of income for the region. Laws protecting Amazon land are hard to enforce, however GPS has made areas not complying more easily detectable. Researchers are analyzing methods for forest regrowth and results are promising. Overfarmed and overstripped abandoned land previously thought useless actually shows regrowth, particularly in areas at the border of forest and pastures where birds spread seeds. Smaller fields therefore, could regrow more quickly than a single large pasture. Guidelines for a most intensive approach developed from this observation could possibly accommodate both developers and environmentalists.

DISCUSSION OF UNIT THEMES

Central themes of this unit are population growth, urbanization, and push/pull factors. Many of Latin America's cities are growing at an extremely fast pace. Most of this urbanization is occurring to accommodate the ever growing population. While part of the population increase can be attributed to immigration from outside the region, much of it is due to internal regional migration. People move based on push and pull factors. Push factors are variables that push people out of the region they currently inhabit and pull factors are those variables that attract and pull people to someplace else. In the case of Latin America, lack of opportunity, unemployment, and land reform measures push people from their current locations. Cities offer a better chance of employment, as well as increased opportunity for education and health resources.

In Mexico many of the residents of the "hollow core" migrate northward for work in the maquiladoras and in Brazil, residents of the poorer northeastern regions are migrating south to Sao Paulo. This migration trend has resulted in extreme population growth and an urban pattern of rich versus poor. A typical Latin America city exhibits a ring of squatter settlements surrounding a more affluent city core. In some cases, such as Guatemala, the rise in population is putting a strain on the local resource base. The available land simply cannot support the growing number of people.

Income and lack of available land has resulted in people encroaching upon areas that were once remote and uninhabited. Brazil loses thousands of acres of rainforest to deforestation. The controversy surrounding the permanent clearing of tropical forests is a prime example of the antagonistic relations that may form when globalization forces and local traditions work against one another. The rainforest land, while home to many of Brazil's indigenous peoples, is now also providing much needed income from cattle herding and logging (*Contemporary World Regional Geography*, Bradshaw/White/Dymond, © 2004, McGraw-Hill Higher Education).

TEST YOUR KNOWLEDGE

Multiple Choice and True/False Questions

1. Why is north central Mexico referred to as the "hollow core?"

    a. There are extremely low population numbers.
    b. There is a large basin in the center of the state.
    c. because of its relative location and its lack of resources
    d. because of its excess commercial activity

2. Which of the following best describes the current population patterns among the Maya?

    a. The Mayan population has remained stable for the past 50 years.
    b. The Mayan population has never rebounded from its decimation during the Spanish conquest.
    c. The Mayan population will most likely die out completely before the year 2050.
    d. The Mayan population has significantly rebounded to dangerous levels.

3. Which of the following best describes Ecuador's topography?

    a. mountainous and volcanic
    b. flat and lowlands
    c. desert
    d. dry savanna

4. Which Andean town lies at the base of Mt. Tungurahua?

    a. Quinto
    b. Buenos Aires
    c. Santiago
    d. Banos

5. In terms of territorial morphology, Chile is considered which shape?

    a. elongated
    b. protruded
    c. compact
    d. fragmented

6. Which of the following has NOT aided Chile in becoming a Pacific Rim player?

    a. varying climates
    b. coastal location
    c. Andean water resources
    d. landlocked location

7. Most of Sao Paulo's poor live:

    a. in center city ghettos.
    b. in rural areas far from city limits.
    c. in shantytowns on the outskirts of the city.
    d. both in shantytowns and elite central areas.

8. The primary economic activities in the Amazon region are:

    a. cattle ranching and logging.
    b. cattle ranching and tourism.
    c. logging and fishing.
    d. fishing and agriculture.

9. True/False Brazil is an ethnically homogenous country.

10. True/False Equal opportunity for work exists in both the cities and rural areas of Guatemala.

Short Answer and Essay Questions

1. Discuss the arguments of both the environmentalists and developers regarding land-use in Amazonia.
2. Why are Sao Paulo's shantytowns expanding?
3. Discuss some the ways in which Chile's elongated shape has helped it become a Pacific Rim player.
4. Discuss the problems researchers face when deciding whether or not to announce a possible volcanic eruption.
5. What impacts have the Mayan population explosion had on land resources?

Map Exercises

1. Place the following physical features on the outline map: the Andes, the Amazon.
2. Using the blank outline map, locate the following: Brazil, Ecuador, Chile, Santiago, Banos.

0 miles 1000
0 kilometers 1000

0 miles 1000
0 kilometers 1000

0 miles 1000
0 kilometers 1000

0 miles 1000
0 kilometers 1000

UNIT TEN: NORTH AMERICA

Unit Objectives

At the completion of the unit, you should be able to:

- understand the process of urbanization, urban sprawl, and edge cities.
- understand cultural landscapes and the importance of cultural identity.
- understand the impact of urbanization on downtowns.

PROGRAM 24

CITYSCAPES, SUBURBAN SPRAWL

Key Themes

- **Urban sprawl**
- **Edge cities**
- **Downtown decline**
- **Downtown revitalization**

Boston

Boston faces the same challenge as many American cities—a declining downtown due to economic and population expansion in the nearby suburbs. Inner city neighborhoods are populated by the unemployed and minorities while the wealthiest move farther away, widening the gap between the rich and poor. In 1994, Boston entered a competition for federal grants earmarked to help some of America's most poverty-stricken areas, or "empowerment zones." GIS was used to establish Boston's empowerment zone by mapping relative poverty levels for each census track. It became clear that defining poverty and which areas were in most need would be a difficult task since so many variables overlap. New enterprise will most likely succeed only if it exploits existing geographic advantages, so the empowerment team incorporated history, economics, ethnicity, as well as spatial elements such as access to transportation, the CBD, and the harbor to develop a plan that won a $47 million second prize. Six years later, progress can be seen in the landscape with building facelifts, remodeling, and new construction.

Suburban Chicago

Intense competition for land in the Chicago area has led to the loss of hundreds of acres of good crop land. Many businesses that were once located downtown are now moving to suburban locations. As businesses relocate, so must their employees and land that hosts some of the country's best soils is being converted from cropland to suburban development. New suburban "edge cities" are being formed as the Chicago metropolitan area pushes farther and farther out into suburban farming communities. Transportation technology has made it possible for downtown commuters to live farther from the city creating a continuous stream of urban sprawl. The "small town" way of life is being threatened along with agricultural land. People want peace and quiet, along with shorter commutes and conveniences of modern life. The high price developers are willing to offer for farmland is often hard to refuse, although there are some who are still holding out in an effort to slow down the pace of development.

PROGRAM 25

ETHNIC FRAGMENTATION IN CANADA

Key Themes

- **Cultural landscape**
- **Landscape change**
- **Ethnic identity**

Montreal

Quebec is a predominately French-speaking province within the English-speaking country of Canada. There has been a growing movement for Quebec independence, however in 1995, the people voted by a slim margin to remain part of Canada. Quebec's French-speaking majority has the lowest birth rate in the province. It is feared that this, coupled with immigration from outside the region, could threaten traditional French culture. Montreal lies at the heart of the issue. Divided into two distinct areas, Montreal has both an English and French side. Many new immigrants choose to settle in the English side, with the opinion that English is the language to learn to be successful in a global economy. The Quebecois however, stress the importance of maintaining French culture in the region and have enacted new laws to protect their language, such as requiring public signs to be written in French. This caused many English-speaking companies to leave in the 70s and 80s. This law has become somewhat relaxed over the years, however the French language is still very much part of the cultural landscape. So much is the importance of the maintaining the French language, that the government pays new immigrants to learn it.

Vancouver

The cultural landscape of Vancouver is being transformed by the influence of its largest minority, East Asians. After Hong Kong was reunified with China, many wealthy Chinese found Vancouver was an obvious new place to settle due to its proximity and its already established Chinatown. The city is now divided with the densely packed east, traditionally the home of non-English speaking immigrants and the wealthier west side where there is more space and more trees. It is the more spacious west side where the new Asian immigrants, accustomed to living in places where land is sparse, have chosen to settle. Houses are now being designed to fit the taste of the new elite and the result is large, "unfriendly" homes which don't blend into the existing cultural landscape—a result of global forces from far beyond the region. Recently new design guidelines have been adopted which require local elements to be incorporated into the new designs. These new design compromises help to bridge the culture gap.

PROGRAM 26

REGIONS AND ECONOMIES

Key Themes

- **Diffusion**
- **Environmental protection**
- **Land rights**

Oregon

The Columbia River in Oregon is the lifeblood for the people and wildlife that live in and around its path. Unfortunately, human development in the area has left the Umatilla tributary very shallow rendering it uninhabitable for the once plentiful salmon. Dams may be the main culprit. Instead of flowing along its usual course, a dam now diverts the water to a canal where it is taken to irrigate fields in the lower valley. Thanks to irrigation, the valley produces millions of dollars of agricultural crops. Unfortunately, there isn't enough water for both the crops and salmon. Debate now ensues between the farmers and local Indians over use of the water. In 1999, taxpayer money was used to complete an engineering project to pump up water from the larger Columbia River. Although a small success for the salmon of the Umatilla, salmon are still threatened on the Columbia because of dams needed to produce electricity for the ever-growing population of the region. Difficult decisions regarding the expanding demand for energy and the protection of wildlife lie ahead.

U.S. Midwest: Spatial Innovations

During the past few decades, Japanese auto manufacturers have challenged North America's lead in the market. In this era of globalization, production techniques have diffused and the primary diffusion path has been from Japan to North America. After the devastation of World War II, Japan was forced to adopt a more streamlined, efficient method of production, relying of small batches and flexibility to keep costs down. Since the market was small, large, grand scale production was not necessary. This "just-in-time" production method produced a better quality car and diffused to the United States when Honda opened a plant in the American Midwest—the most advantageous location for minimizing shipping costs. Although great at producing quality cars, "just-in-time" production limits profits. At times, so much quality was built into a car that it became unaffordable for the consumer. Now auto manufacturers in Japan and North America alike are adapting production methods to produce a high quality car that consumers are willing to pay for.

DISCUSSION OF UNIT THEMES

The predominantly urbanized nature of the population of the United States resulted in many very large cities and groups of cities. By the 1990s, large and growing metropolitan areas were increasingly dominant in the U.S. human geography. Urbanization within the United States produced distinctive landscape forms through the types of buildings, the differentiation of land uses, and the distribution of people. From the late 1970s, expanding American urbanism spread new land uses into former rural areas. The term "edge city" was coined for new exurban developments relying primarily on car and truck transport and secondarily on air travel. Edge cities include large developments of shopping malls, offices, warehouses, and factories, located on the edge of major metropolitan areas. Since the 1970s, there were some movements of people from the suburbs farther out from the city into surrounding rural areas, often encroaching upon farming areas as was seen in Chicago. In many cases, this process leads to the decline of the downtown. Cities then must decide if and how they want to revitalize them (*Contemporary World Regional Geography*, Bradshaw/White/Dymond, © 2004, McGraw-Hill Higher Education).

Gentrification involves attracting men and women in a thirty-something age group into downtown areas. The movements of people with high incomes into city centers was seen as the solution, however the resulting increase in land prices often forced lower income residents to move elsewhere. Cities such as Boston have chosen to declare empowerment zones and apply for federal funding to rebuild their downtown (*Contemporary World Regional Geography*, Bradshaw/White/Dymond, © 2004, McGraw-Hill Higher Education).

The cultural landscape is the human imprint on the physical environment and it is often best seen in cities. Cities such as Montreal and Vancouver have a distinctive cultural landscape, which embodies their rich diverse ethnic character. French culture is obvious in the Montreal landscape and Asian culture thrives in Vancouver. While this mix provides for an interesting landscape, it can also lead to conflict when one culture wants to dominate. Cultural landscapes are linked with identity and historical residents can become threatened when their cultural environment changes. Governments often get involved to ensure a happy medium between cultures (*Contemporary World Regional Geography*, Bradshaw/White/Dymond, © 2004, McGraw-Hill Higher Education).

TEST YOUR KNOWLEDGE

Multiple Choice and True/False Questions

1. Which of the following Canadian provinces contains the largest French population?

   a. Alberta
   b. British Columbia
   c. Ontario
   d. Quebec

2. Vancouver's cultural landscape is changing because of influence of immigrants from which region?

   a. South Asia
   b. East Asia
   c. Southeast Asia
   d. Australia

3. Which of the following best describes an empowerment zone?

   a. a special area of the worst poverty
   b. the richest area of a city
   c. a suburban edge city
   d. a poor rural area

4. Which of the following best describes "Just-In-Time" production?

   a. a control system that is purely management driven
   b. a control system that aims to produce the most in the shortest period of time
   c. a control system that takes its cues not from a manager, but from the workers at the end of the line—the point of final assembly
   d. none of the above

5. Which country first utilized "Just-In-Time" production methods?

   a. Japan
   b. China
   c. Canada
   d. United States

6. Which resource is under debate in Oregon?

    a. timber
    b. water
    c. gold
    d. oil

7. The largest city in Quebec is:

    a. Montreal.
    b. Toronto.
    c. Regina.
    d. Quebec City.

8. True/False Edge cities usually form at least 500 miles from a major metropolitan area.

9. True/False The province of Quebec is predominantly French.

10. True/False Chicago is losing good agricultural land to urbanization.

Short Answer and Essay Questions

1. What are some of the ways in which Quebec is attempting to preserve the French language?
2. Describe the debate over water resources along the Columbia River in Oregon.
3. What processes are converting farmland to developed land in the Chicago area?
4. Describe the changes taking place in Vancouver's cultural landscape.
5. Discuss some ways cities are trying to revitalize their downtowns.

Map Exercises

1. Place the following physical features on the outline map: the Columbia River, the Mississippi River, and the Rocky Mountains.
2. Using the blank outline map, make a dot for the location of the following North American cities: Vancouver, Boston, Montreal, and Chicago.

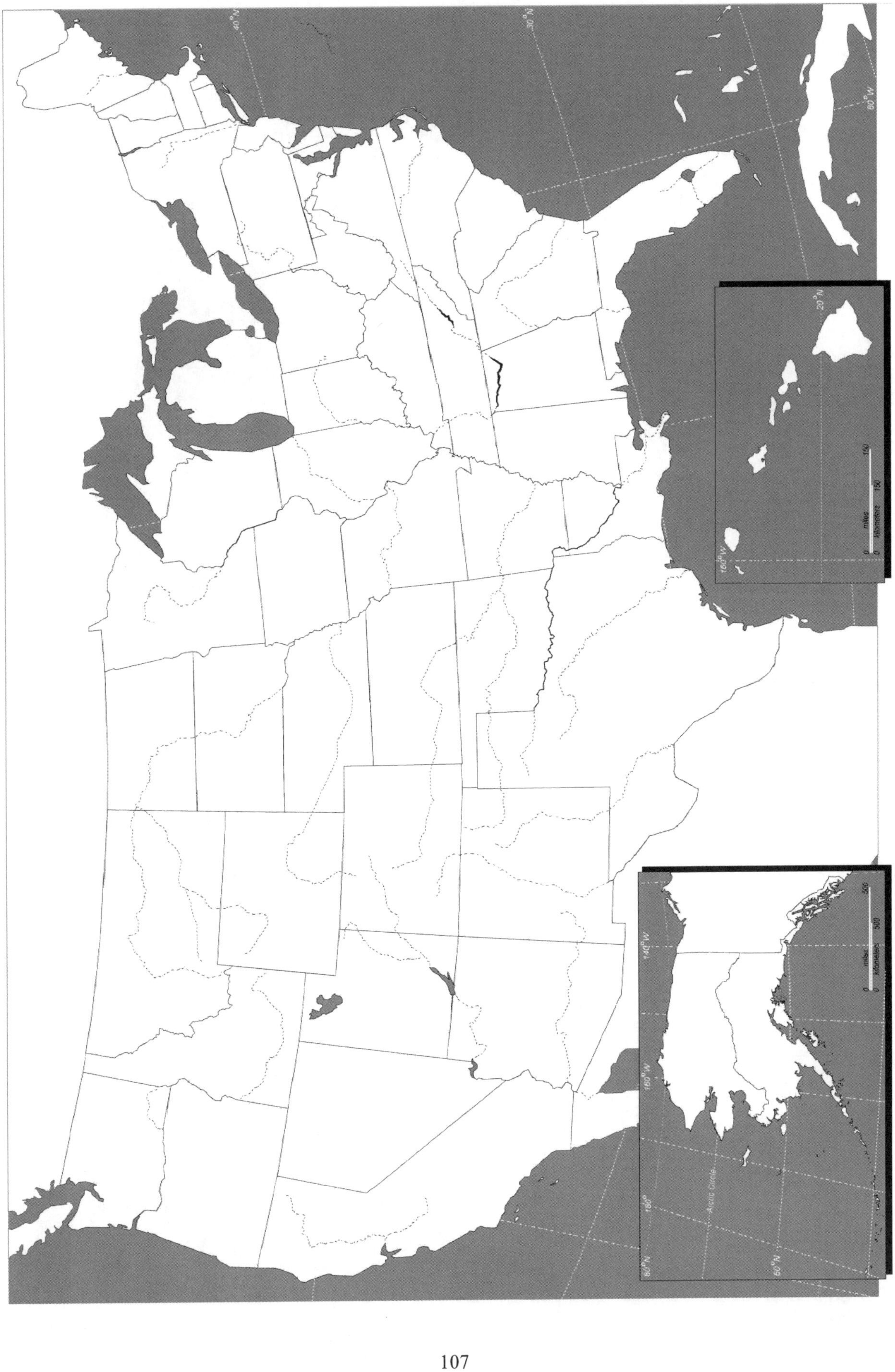
30°N
80°W
20°N
160°W
miles
kilometers
150
140°W
500
Arctic Circle
180°
60°N

40°N
30°N
80°W
20°N
160°W
miles
kilometers
150
150
140°W
160°W
500
500
Arctic Circle
180°
60°N
80°N

miles
kilometers

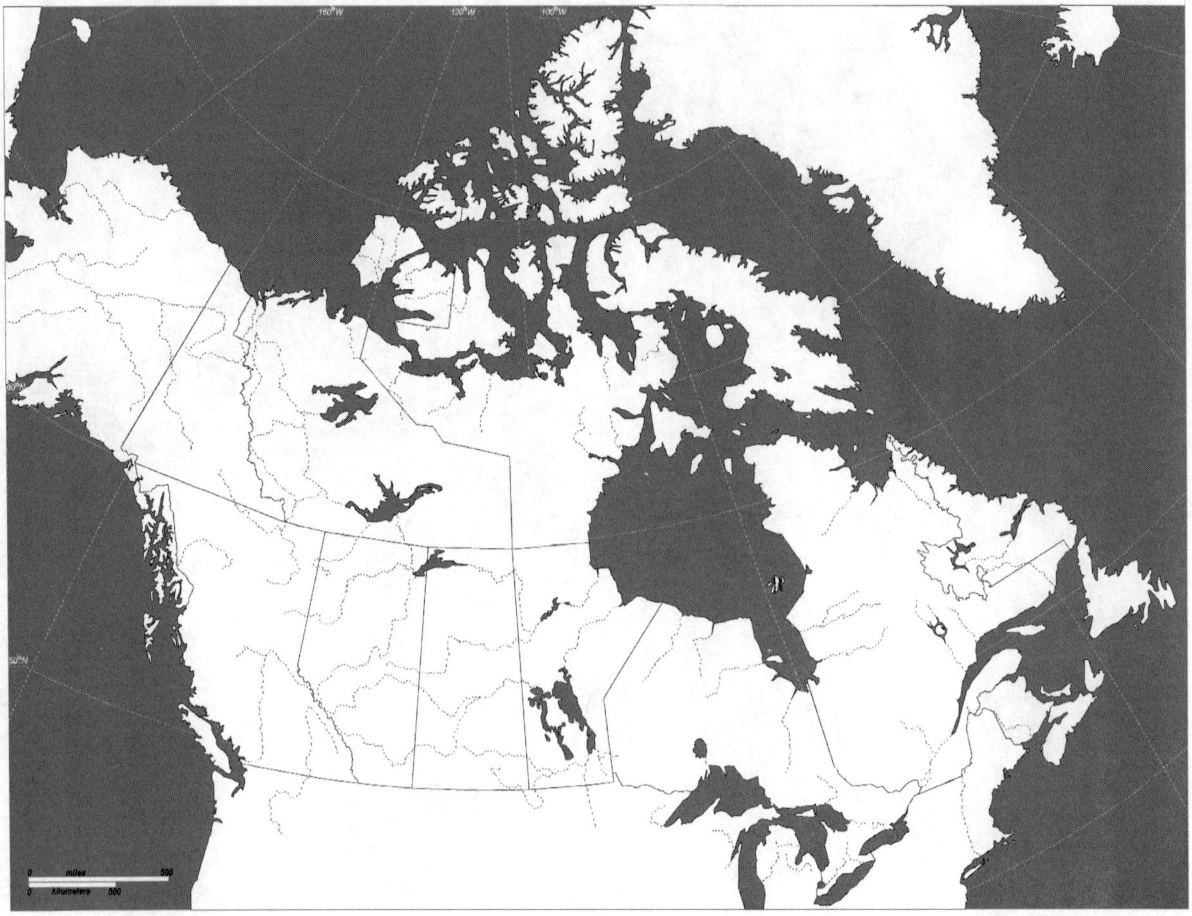
160°W
120°W
100°W
0 miles 500
0 kilometers 500

160°W
120°W
100°W
0 miles 500
0 500

ANSWER KEY

UNIT ONE

1. a
2. c
3. b
4. a
5. d
6. a
7. d
8. b
9. f
10. f

UNIT TWO

1. a
2. b
3. d
4. b
5. a
6. a
7. a
8. b
9. b
10. c
11. f
12. f

UNIT THREE

1. a
2. d
3. b
4. b
5. d
6. c
7. b
8. f
9. f
10. f

UNIT FOUR

1. b
2. a
3. d
4. d
5. c
6. d
7. f
8. f
9. f
10. t

UNIT FIVE

1. a
2. b
3. c
4. a
5. a
6. a
7. c
8. b
9. t
10. t

UNIT SIX

1. c
2. a
3. a
4. f
5. f

UNIT SEVEN

1. c
2. b
3. a
4. a
5. c
6. a
7. t
8. t
9. t
10. t

UNIT EIGHT

1. b
2. a
3. a
4. b
5. d
6. d
7. f
8. t
9. t
10. f

UNIT NINE

1. c
2. d
3. a
4. d
5. a
6. d
7. c
8. a
9. f
10. f

UNIT TEN

1. d
2. b
3. a
4. c
5. a
6. b
7. a
8. f
9. t
10. t